T0259987

MATH!

Books by Serge Lang of Interest for High Schools

Geometry (with Gene Murrow)

This high school text, inspired by the work and educational interests of a prominent research mathematician, and Gene Murrow's experience as a high school teacher, presents geometry in an exemplary and, to the student, accessible and attractive form. The book emphasizes both the intellectually stimulating parts of geometry and routine arguments or computations in physical and classical cases.

MATH! Encounters with High School Students

This book is a faithful record of dialogues between Lang and high school students, covering some of the topics in the *Geometry* book, and others at the same mathematical level. These encounters have been transcribed from tapes, and are thus true, authentic, and alive.

Basic Mathematics

This book provides the student with the basic mathematical background necessary for college students. It can be used as a high school text, or for a college precalculus course.

The Beauty of Doing Mathematics

Here, we have dialogues between Lang and audiences at a science museum in Paris. The audience consisted of many types of persons, including some high school students. The topics covered are treated at a level understandable by a lay public, but were selected to put people in contact with some more advanced research mathematics which could be expressed in broadly understandable terms.

First Course in Calculus

This is a standard text in calculus. There are many worked out examples and problems.

Introduction to Linear Algebra

Although this text is used often after a first course in calculus, it could also be used at an earlier level, to give an introduction to vectors and matrices, and their basic properties.

Serge Lang

MATH!

Encounters with High School Students

With 103 Illustrations

Springer-Verlag
New York Berlin Heidelberg London Paris
Tokyo Hong Kong Barcelona Budapest

Serge Lang
Department of Mathematics
Yale University
New Haven, CT 06520
U.S.A.

Mathematics Subject Classifications (1991): 00A05, 00A06

Cover photo courtesy of Carol MacPherson.

Library of Congress Cataloging in Publication Data
Lang, Serge
Math! : encounters with high school students.
1. Mathematics–Addresses, essays, lectures.
I. Title.
QA7.L284 1985 510 85-13837

Printed on acid-free paper.

A French version was published as: *Serge Lang, des Jeunes et des Maths*, Belin, 1984.

Typeset by House of Equations, Inc., Newton, New Jersey.
Printed and bound by Braun-Brumfield, Ann Arbor, MI.
Printed in the United States of America.

9 8 7 6 5 4 3 2

ISBN 0-387-96129-1 Springer-Verlag New York Berlin Heidelberg Tokyo
ISBN 3-540-96129-1 Springer-Verlag Berlin Heidelberg New York Tokyo

Acknowledgement

I want to thank the teachers who made it possible for me to meet their classes: William Bisset, Patricia Chwat, Peter Edwards, Marie-Therese Giacomo, Michel Ricart. I know that there are some bad teachers, but there are also some good ones, and the students with whom I did mathematics would certainly not have reacted as they did if their teachers had been bad. I thank Abe Shenitzer, who induced me to give the talks in Canada, and organized them. I also thank Jean Brette, who directs the Mathematics Section of the Palais de la Découverte in Paris, for the contacts which arose through him with high school teachers in Paris, and also for his interest in the whole pedagogical enterprise. I also appreciated Stephane Brette's interest, and his willingness to take part in a mathematical dialogue with me, after one of the talks. I thank Patrick Huet who video-taped for the Paris IREM two of the talks, and the discussion reproduced at the end of this book. Finally I thank Carol MacPherson for the photograph on the cover.

SERGE LANG

Contents

Who is Serge Lang?

Serge Lang was born in Paris in 1927. He went to school until the 10th grade in the suburbs of Paris, where he lived. Then he moved to the United States. He did two years of high school in California, then entered the California Institute of Technology (Caltech), from which he graduated in 1946. After a year and a half in the American army, he went to Princeton in the Philosophy Department where he spent a year. He then switched to mathematics, also at Princeton, and received his PhD in 1951. He taught at the university and spent a year at the Institute for Advanced Study, which is also in Princeton.

Then he got into more regular positions: Instructor at the University of Chicago, 1953–1955; Professor at Columbia University, 1955–1970. In between, he spent a year as a Fulbright scholar in Paris in 1958.

He left Columbia in 1970. He was Visiting Professor at Princeton in 1970–1971, and Harvard in 1971–1972. Since 1972 he has been a professor at Yale.

Besides math, he mostly likes music. During different periods of his life, he played the piano and the lute.

From 1966 to 1969, Serge Lang was politically and socially active, during a period when the United States faced numerous problems which affected the universities very deeply.

He has also been concerned with the problems of financing the universities, and of their intellectual freedom, threatened by political and bureaucratic interference. As he says, such problems are invariant under ism transformations: socialism, communism, capitalism, or any other ism in the ology.

However, his principal interest has always been for mathematics. He has published 28 books and more than 60 research articles. He received the Cole Prize in the U.S. and Prix Carriere in France.

Dear Christopher, Rachel, Sylvain, Yaelle, and all the others

I am writing this to you because I came to schools like yours, in France as well as in Canada, to talk mathematics with students who could have been your own friends, and who thus contributed to a joint enterprise.

I wanted to show them beautiful mathematics, at the level of your class, but conceived the way a mathematician does it. In most school books, the topics are usually treated in a way which I find incoherent. They pile up one little thing on another, without rhyme or reason. They accumulate technical details endlessly, without showing the great lines of thought in which technique can be inserted, so that it becomes both appealing and meaningful. They don't show the great mathematical lines, similar to musical lines in a great piece of music. And it's a great pity, because to do mathematics is a lively and beautiful activity.

This book is made up of several lectures, or rather dialogues, which have been transcribed as faithfully as possible from the tapes, to preserve their lively style. It gave me great pleasure to have this kind of exchange with all the students, in different classes. The subjects concern geometric and algebraic topics, understandable at the ninth and tenth grade level. I even gave one of the talks to an eighth grade class! If students at those levels could understand and enjoy the mathematics involved, so can you.

Each dialogue is self-contained, so you don't have to read this book continuously from beginning to end. Each topic forms a single unit, which you can enjoy independently of the others. If, while reading any one of them, you find the going too rough, don't let that put you off. Keep on reading, skip those passages that don't sink in right away, and you will probably find something later in the lecture which is easier and more accessible to you. If you are still interested, come back to those passages which gave you trouble. You will be surprised how often, after sleeping on it, something which appeared hard suddenly becomes easy. Just browse through the book, pick and choose, and mostly get your mind to function.

A lot of the curriculum of elementary and high schools is very dry. You may never have had the chance to see what beautiful mathematics is like. I hope that if you are a high school student, you will be able to complement whatever math course you are taking by reading through this book.* I have many objections to the high school curriculum. Perhaps the main one is the incoherence of what is done there, the lack of sweep, the little exercises that don't mean anything. You will find something quite different here, which I hope will inspire you. And by doing mathematics, you might end up by liking math as you like music, or as I like it.

SERGE LANG

---

* The first five talks would fit well in a geometry course, and the last two in an algebra course. See also the book which I wrote in collaboration with Gene Murrow: *Geometry*, published by Springer-Verlag.

What is pi?

The following talk was given at a high school in the suburbs of Toronto, April 1982, to a class of students about 15 years old. The talk lasted about 1 hour and 15 minutes.

SERGE LANG. My name is Serge Lang, I usually teach at Yale, but today I came here to do mathematics with you.

We are going to study the area of some simple geometric objects, like rectangles, triangles, and circles which you must have heard about in this course. Let's start with rectangles. We assume that its area is the product of the base times the height, so if the sides have lengths a and b, then the area of the rectangle is ab. For instance, if a rectangle has sides of lengths 3 cm and 4 cm, then its area is 12 cm^2. You can check this on the figure.

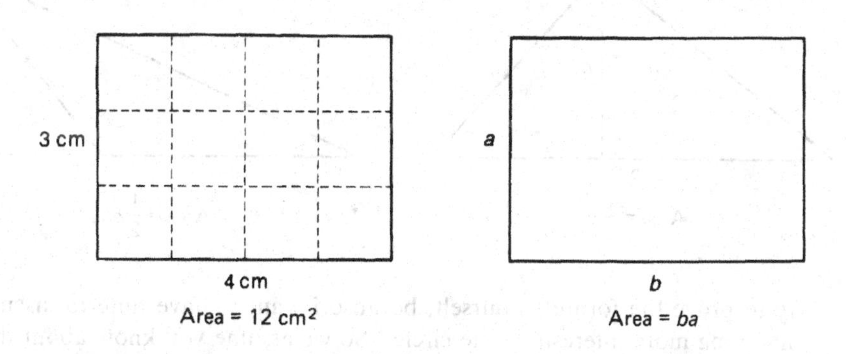

If you cut a rectangle in half, like this:

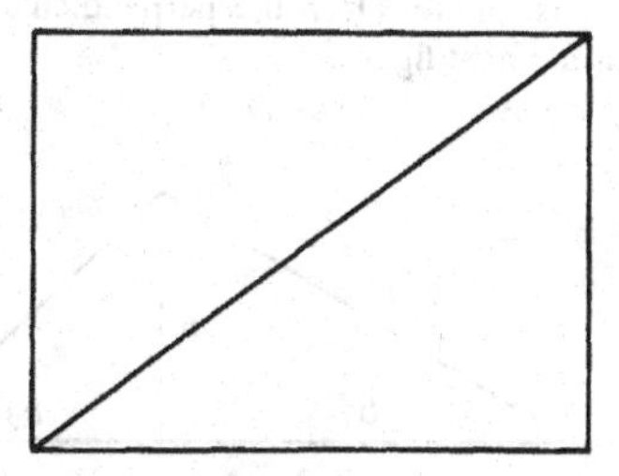

then you get a right triangle, so the area of the right triangle is one half the product of the base times the height. We can write

$$\text{area of right triangle} = \frac{1}{2}bh$$

where b is the base and h is the height.

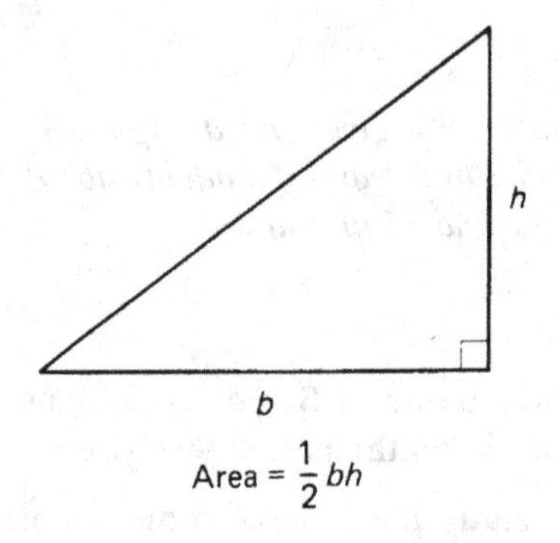

You should also know that this formula is true for any triangle, if h is the perpendicular height. I can show you this on two possible figures:

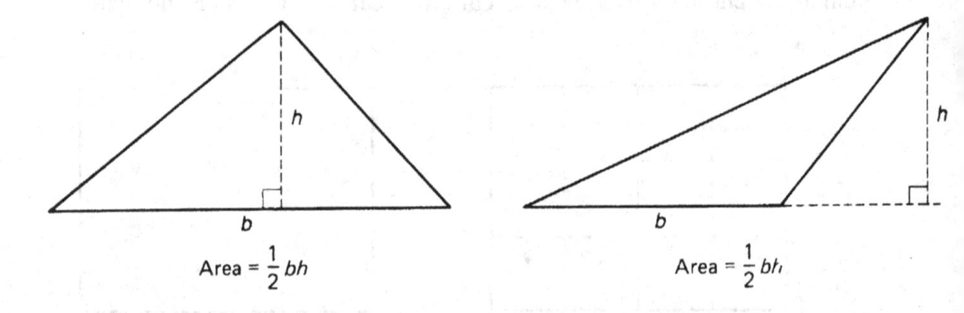

Try to prove the formula yourself, because I want to have time to discuss something more interesting, the circle.[1] So we assume you know about the area of a triangle. Now you have the circle radius r.

[1] I give the proof for the first figure. Drop the perpendicular height from one vertex to the opposite side as shown on the next figure.

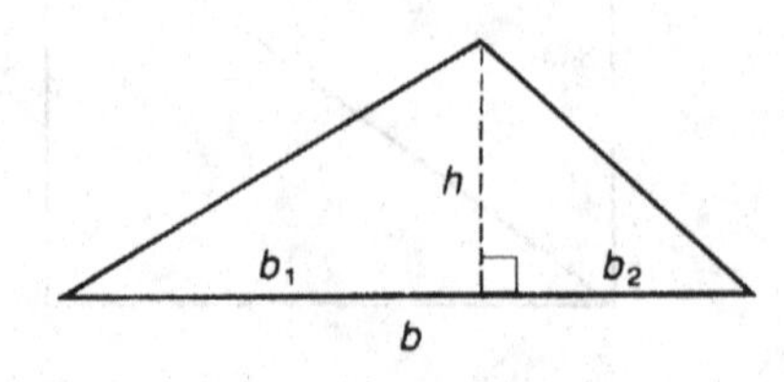

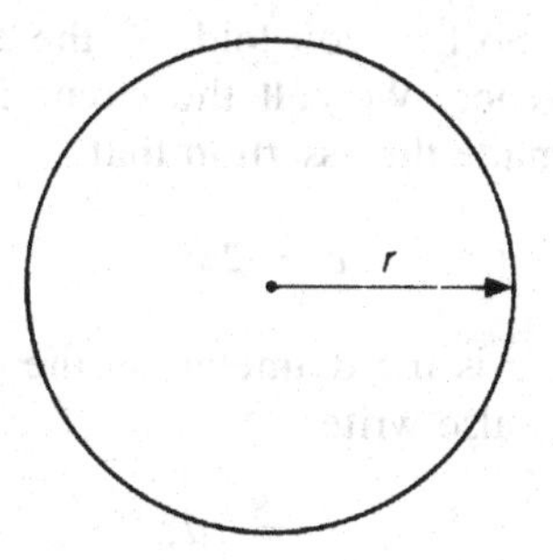

Do you know what is the formula for its area?

A STUDENT. It's pi r squared.

SERGE LANG. That's right, it's πr^2. Well, what is pi?

A STUDENT. What is pi?

SERGE LANG. Yes.

THE STUDENT. 3.14

SERGE LANG. You claim that π is 3.14. Is that an exact expression?

THE STUDENT. No, I don't think so.

SERGE LANG. Then why did you say 3.14?

STUDENT. Well, it goes on and on.

SERGE LANG. OK. So we put dot dot dot like this, 3.14... to signify that it goes on and on and on. How do you know how it goes on?

[*The students react variously.*]

SERGE LANG. It's not so clear! So it's a problem how it goes on. I mean, how are you going to compute it?

A STUDENT. You measure the perimeter of the circle and divide by twice the radius.

SERGE LANG. Ah, now you're telling me something else. Instead of the perimeter, let me call it the circumference. Do you mind if I call it the circumference?

STUDENT. No.

Then the triangle is decomposed into two right triangles, whose bases are b_1 and b_2 such that $b_1 + b_2 = b$. The two right triangles have the same height h. Then using the formula for the area of a right triangle, we now get:

$$\text{area of the triangle} = \frac{1}{2}b_1h + \frac{1}{2}b_2h = \frac{1}{2}(b_1 + b_2)h = \frac{1}{2}bh.$$

This proves the formula for the first figure. Treat the second figure similarly. You will need a subtraction instead of an addition.

SERGE LANG. OK. So first you told me the area is πr^2, and now you mention the circumference. We call the circumference c. And what did you just tell me? You made the assertion that

$$c = 2\pi r.$$

That's what you said: $2r$ is the diameter, so the circumference is pi times the diameter, so we can also write

$$c = \pi d,$$

where d is the diameter, $d = 2r$. But now look. You have two formulas, for the area and for the circumference:

$$\pi r^2 \quad \text{and} \quad 2\pi r.$$

By the way, what's your name?

STUDENT. Serge.

SERGE LANG. Oh Serge, just like me! [*Laughter.*] Serge said, to compute π, you look at the circumference and divide by the diameter. The circumference is something you can measure. You can get a soft tape at home, you put it around a frying pan, and you measure the circumference. Then you measure the diameter with a ruler, and divide. Actually you can get one or two decimals accuracy out of that, probably you can get two decimals if you are careful. You get some sort of value, which is an approximation for π.

You would have a much harder time trying to measure the area to get an approximate value for π.

Now the question is: you've got these two formulas, one for the area, one for the circumference. How do you know these formulas are true?

STUDENTS. [*Silence, questioning looks.*]

SERGE LANG. How do you prove them? Has anybody ever broached the problem of proving these formulas? At any time? You were just given the formulas.

STUDENTS. [*Negative looks on most faces, one or two raise their hands.*]

SERGE. You can just say that π equals the circumference divided by the diameter, and work it out.

SERGE LANG. Work what out? You just repeated one of the two formulas. You have two formulas. Suppose I want to prove them. To prove them I have to start from something and then I have to get to the formulas by logic. So I start from what?

SERGE. You start from where you divide the circumference by the diameter, that equals π.

SERGE LANG. And then what? Now you have to reach the area. What is the definition of π? Before you can prove something, you must have a definition.

SERGE. It's what I said, the circumference divided by the diameter.

SERGE LANG. But then, you have to show that it's the same π in the formula for the area. If you tell me that π is the circumference divided by the diameter, which is twice the radius, you can start with that as a definition, but then you have to prove something, which is the other formula.[2]

So we have to start from something, with a definition, otherwise I can't prove anything. And then logically, derive the formulas. So the question is, where do we start from? That's what I am after. I want to start from somewhere, and get to these two formulas.

I will have to explain two things about these formulas. One is where the r^2 comes from; and second, where the π comes from. They come from two different aspects of the problem. One of the aspects has to do with the r^2. Why is there an r^2 in the formulas for the area? And why is there an r (but not r^2) for the circumference? The presence of the r and r^2 has to be discussed. And the other thing I have to dicuss is the π.

So we start all over. I'll first explain the r^2, and after that I'll explain the π.

Let's go back to the even simpler case of the rectangle. Suppose I have a rectangle of sides a, b. Then the area of the rectangle is just the product, ab. Now suppose I take a rectangle whose sides are twice a and twice b, so I blow up the rectangle by a factor of 2. How does the area change?

A STUDENT. It doubles.

SERGE LANG. What's your name?

THE STUDENT. Adolph.

SERGE LANG. The area doubles? What is the area of the new rectangle?

[*Another student starts talking.*]

SERGE LANG. No. Adolph, I am asking Adolph. The area of a rectangle is the product of the sides. Right?

ADOLPH. Yes.

SERGE LANG. So one side of the new rectangle is $2a$, and what is the other side?

ADOLPH. It's $2b$.

---

[2] You also have to prove that no matter what circle you take, the ratio of the circumference by the diameter gives the same number. This is precisely one of the things we are trying to prove. I was not on the ball when I did not raise this objection explicitly that way.

SERGE LANG. That's right, so the total area is $2a$ times $2b$, which is $4ab$.

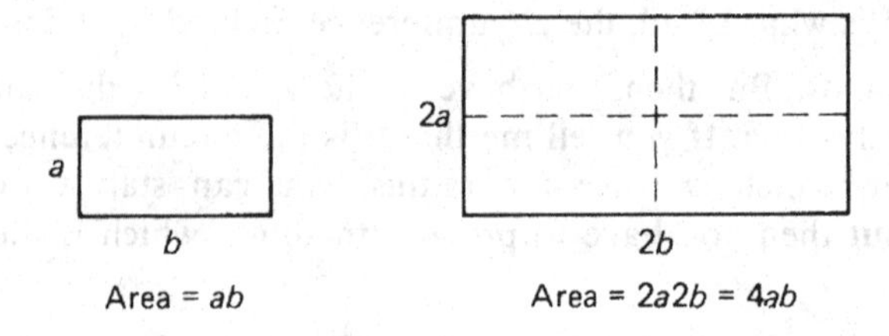

Now suppose I take a rectangle with three times the sides. So here I have sides of lengths $3a$ and $3b$.

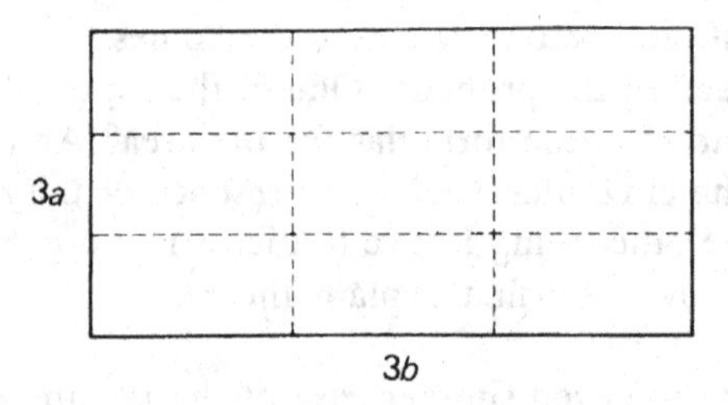

What is the area of the rectangle with three times the sides? Adolph.

ADOLPH. $9ab$.

SERGE LANG. That's right, $9ab$, it's $3a$ times $3b$, which is $9ab$. Suppose I now take a rectangle with one half the sides, so I have here $\frac{1}{2}a$, and $\frac{1}{2}b$. What is the area of this rectangle, with one half the sides?

$\frac{1}{2}b$ $\frac{1}{2}a$

ADOLPH. It's ab over 4.

SERGE LANG. That's right. One fourth ab. Now suppose in general I take a rectangle with the sides ra and rb. Like this.

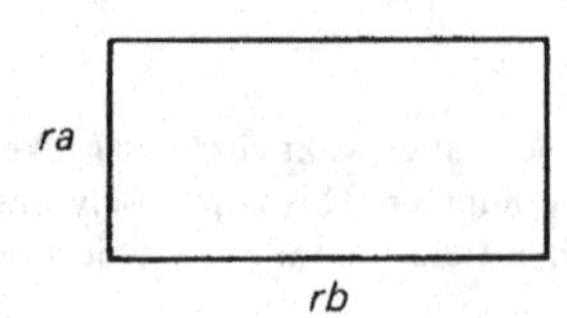

Adolph, what's the area of this rectangle?

ADOLPH. ra times rb.

SERGE LANG. That's right, ra times rb, which is what?

ADOLPH. r squared times ab.

SERGE LANG. Yes, r^2ab. So if I change the rectangle by a blow up by a factor of r, how does the area change? Adolph.

ADOLPH. Can you repeat that, please?

SERGE LANG. Yes. I have my old rectangle. I blow it up by a factor of r. In both directions. You see, the new sides are ra, rb? How does the area change? The old area was ab. What is the new area?

ADOLPH. r squared ab.

SERGE LANG. Yes, r^2ab, not rab. The area changes, by what factor?

ADOLPH. By r^2.

SERGE LANG. You said r a minute ago. Well it's not r. It's r^2. You see it? Does everybody see it?

[*Students agree that they see it.*]

SERGE LANG. So if I make a blow up by a factor of r, the area of a rectangle changes by a factor of r^2. And of course, r can be bigger than 1, or r can be smaller than 1, like $r = 1/2$ or $r = 1/3$. So now you see how area changes for rectangles. Any questions? Everybody got that?

[*No questions.*]

All right, now instead of a rectangle, suppose I have another figure. Suppose I have a curved figure, like this.

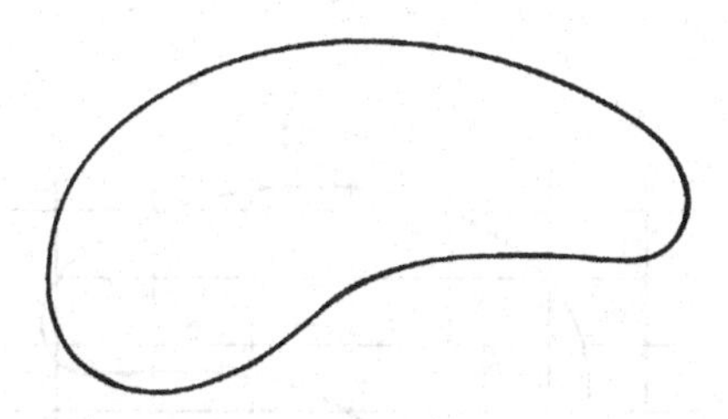

Just like a kidney. Suppose I have a kidney, and it has a certain area A. Now I blow up the kidney by a factor of 2, for example. What will be the area of the blown up kidney?

A STUDENT. A^2?

SERGE LANG. No, let's go back. I have a rectangle with area A. I blow this rectangle up by a factor of 2. What is the new area?

A STUDENT. It's $4A$.

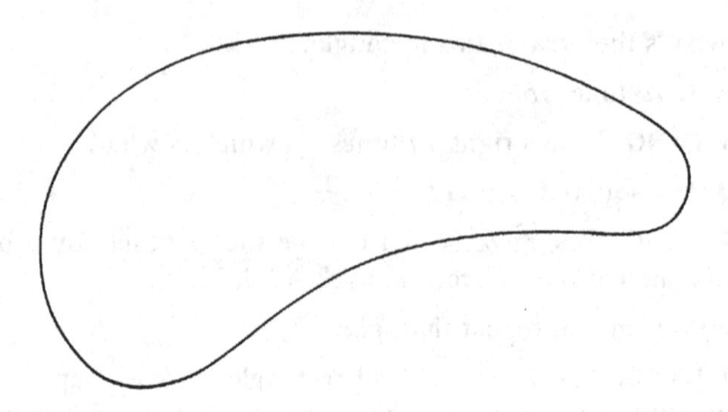

SERGE LANG. Yes, the area changes by 2 squared. If I blow up the rectangle by a factor of r, the area changes by a factor of r^2. Suppose I blow up a curved figure. I blow up by a factor of 2. The area will change by what factor?

A STUDENT. 4.

SERGE LANG. OK, the area changes by this same factor of 4. Why? What is the proof? I have a curved figure, a kidney, not a rectangle. How do I prove it? Anybody have any ideas? All right, Adolph.

ADOLPH. You measure around.

SERGE LANG. No, you don't determine area by measuring around. If I measure around, I get the perimeter, the circumference. I am now dealing with the area. The whole thing inside.

[*A student raises her hand. There is a certain amount of fumbling around by several students. After a while, Serge Lang picks up again.*]

SERGE LANG. I try to reduce the question to rectangles. I make a grid like this.

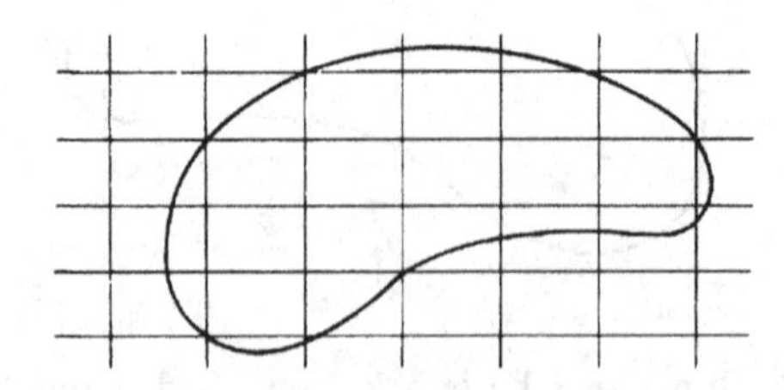

You see the grid? Then the area of the kidney is approximated by the area of the rectangles which are inside the kidney. I look at all the rectangles here which lie completely inside the kidney. [*Serge Lang draws the thick line in the next figure.*] They go like that, all the way down there, up here, there, and here.

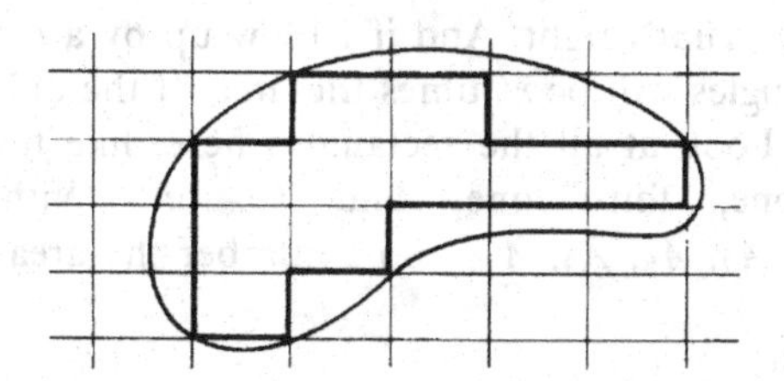

So if I take the rectangles here, inside the thick line, I get an approximation of the area of the kidney by rectangles. If I make the grid very fine, as in the next picture, I get an even better approximation.

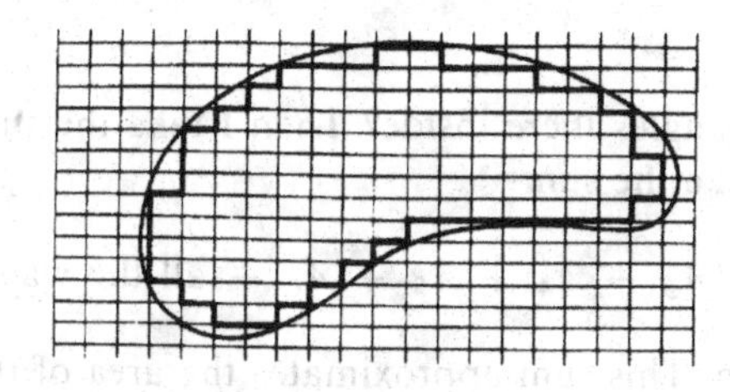

You see all these little rectangles there inside? If I take the sum of these little rectangles, the sum of the area of these little rectangles, then I get a good approximation of the area of the kidney.

Now I blow up the whole picture by a certain factor r. We make a dilation, that is an expansion, or a contraction, by a certain factor r. Then the rectangles also dilate by a factor of r. For instance, if r is greater than 1, then a rectangle in the first drawing gets blown up to a bigger rectangle.

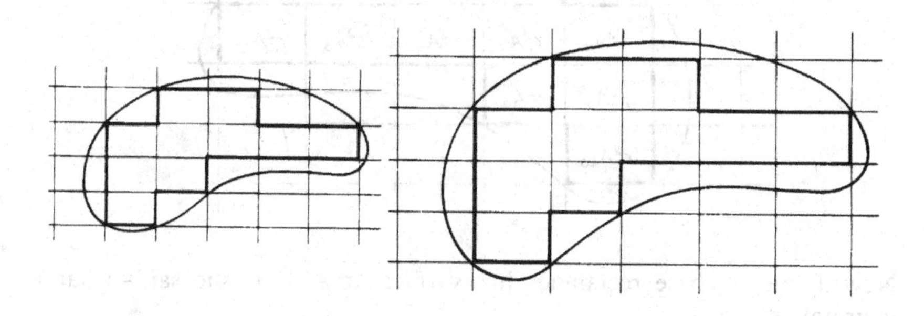

A STUDENT. Yes, if you blow up these rectangles inside the kidney by a factor of 2, then the area of the blown up rectangles will be four times the old area.

SERGE LANG. That's right. And if I blow up by a factor of r, the area of the new rectangles will be r^2 times the area of the old rectangle. Everybody sees that? Look at all the rectangles here, like this, there is a first one, second one, third one, and so on. With areas A_1, A_2, A_3, A_4, A_5, A_6, A_7, A_8, A_9, A_{10}. So I number the areas of the little rectangles.

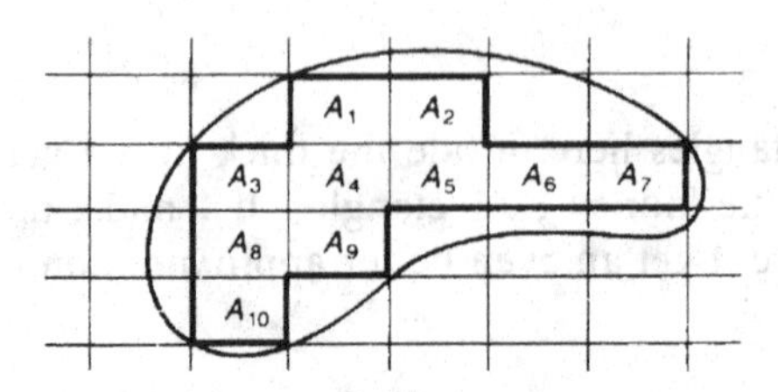

You see these rectangles there inside? Then I take the sum of the areas of these rectangles, like the sum

$$A_1 + A_2 + A_3 + A_4 + A_5 + A_6 + \text{all the way up to } A_{10}.$$

I have ten of them. This sum approximates the area of the kidney. Right? Now I make a blow up of the whole situation, like in photography by a factor–of whatever you want. What do you want: 2 or do you want r? Do you want a specific number, or can I use r?

SEVERAL STUDENTS. r. You can use r.

SERGE LANG. I can use r. OK. So I make a blow up by a factor of r.

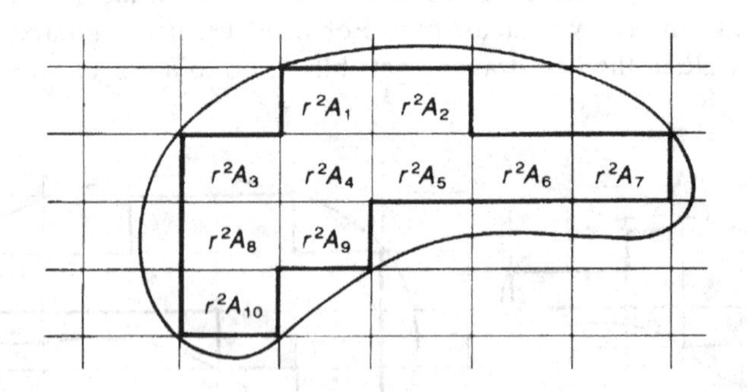

Now if one of these rectangles has surface area A, as she said–what is your name?

STUDENT. Rachel.

SERGE LANG. Rachel. As Rachel said, the area of the dilated rectangle will be r^2A. So if I have here a rectangle with area A, over there I will have a rectangle with area r^2A. So what will be the sum of all the blown up rectangles? It will be

$$r^2A_1 + r^2A_2 + r^2A_3 + \text{up to } r^2A_{10}.$$

RACHEL. Yes.

SERGE LANG. If I factor out r^2, I get r^2 in front, so I get

$$r^2(A_1 + A_2 + \text{up to } A_{10}).$$

But $A_1 + A_2 + \cdots + A_{10}$ is an approximation of the area of the kidney, and

$$r^2(A_1 + A_2 + \cdots + A_{10})$$

is an approximation of the area of the kidney dilated by r, so it is an approximation of the new area, that is r^2 times the old area.

Now if I make a very fine grid, the area of the rectangles inside the kidney gets closer and closer to the area of the kidney. And that's the reason why the area of the kidney also changes by a factor of r^2. Do you understand the argument? Not quite?

A STUDENT. Kind of.

SERGE LANG. Kind of . . . Did you understand the argument about the rectangles? If I blow up each rectangle by a factor of r, its area changes by a factor of r^2. Do you see that I can approximate the curved figure by rectangles? How do I approximate the curved figure by rectangles?

RACHEL. By adding up the area of the little rectangles.

SERGE LANG. That's right. I first make a grid and then I add up all the areas of the little rectangles. And that gives me a good approximation of the area of the kidney. Then I blow up the whole picture by a factor of r. Each little rectangle gets blown up by a factor of r, and the area of each rectangles changes by a factor of r^2, so their sum also changes by a factor of r^2. And so I claim that is an argument why the area of the curved figure will also change by a factor of r^2. Do you accept that? Any comments?

[*Various students discuss the question.*]

ONE STUDENT. It gets bigger?

SERGE LANG. Yes, if r is bigger than 1, it gets bigger. If r is 1/2 it will shrink by a factor of 1/4th. That's right. Do you accept it? Anybody else?

ANOTHER STUDENT. Yeah, I accept it. [*Laughter, the statement sounded funny.*]

SERGE LANG. Do you have any objections to the argument?

A STUDENT. I have a question. If r was less than 1, would it reduce by a factor r to the negative 2?

SERGE LANG. No. If r is less than 1 then you would get the blow up by a factor of r^2 because r^2 would be less than 1. You can write r as some other number s to the negative one, that is $r = s^{-1}$, so r^2 is s to the negative 2, that is $r^2 = s^{-2}$. Multiplying by r^2 is then really a shrink down, by a factor less than 1.

A STUDENT. So it's not really a blow up.

SERGE LANG. All right, let's use a neutral name, let's call it a dilation, that's a sort of neutral name. If r is like 2 or 3, the dilation increases the area. If r is 1/2, then the area changes by a factor of 1/4th, which is a shrink down. So we use the neutral name, dilation.

Our argument was by approximating the curved figure by rectangles. We can say that under a dilation by a factor of r, area changes by a factor of r^2. Can we go on from that? Can I erase? [*Students nod.*]

We go on. Now we get back to the circle, which is certainly much better than the kidney. I mean, it's curved too, but it's better than the kidney because the curve is more regular.

Take a circle of radius 1. How do you get a circle of radius 2 from the circle of radius 1?

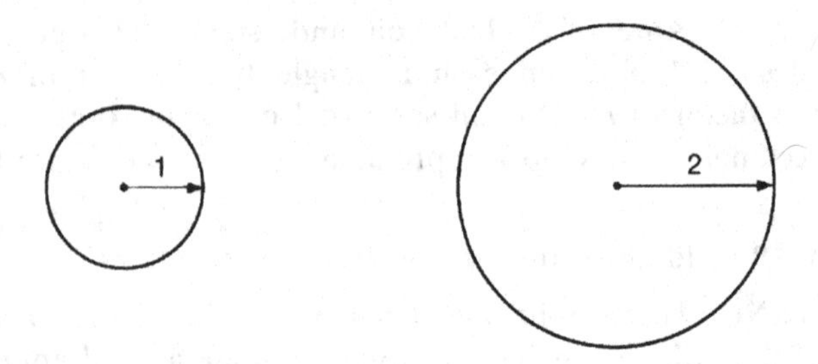

A STUDENT. You increase the radius by 1.

SERGE LANG. By a factor of 1 or a factor of 2? Take another radius. How do you get a circle of radius 10 from the circle of radius 1?

STUDENT. You've increased it by a factor of 10.

SERGE LANG. I've increased it by a *factor* of 10. And in general, if I have a circle of radius r, how do I get the circle of radius r from the circle of radius 1?

STUDENT. You increase it by a factor of radius r.

SERGE LANG. By a factor of r. Yeh? So I can say that a circle of radius r is the dilation of a circle of radius 1 by a factor of r. Do you accept that?

STUDENT. Yeh.

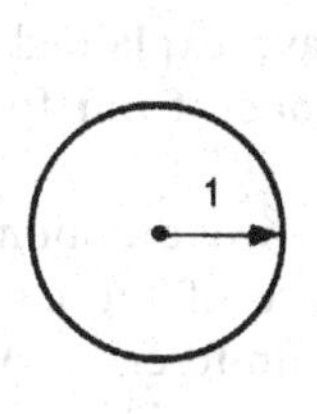

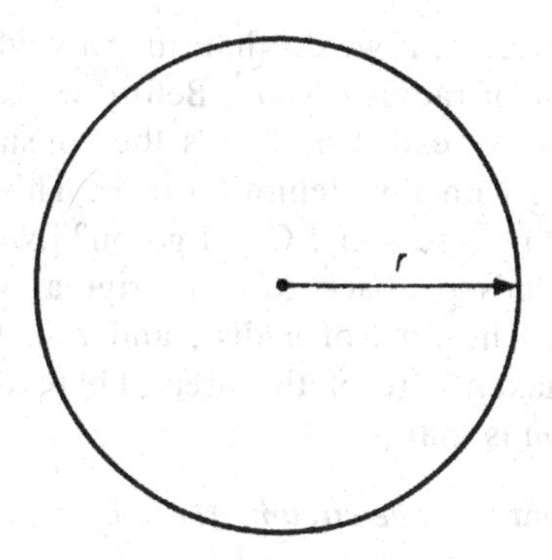

SERGE LANG. And the circle of radius 1/2 is the dilation of the circle of radius 1 by a factor 1/2. Now suppose you know the area of a circle of radius 1? What will be the area of a circle of radius 2? Let A be the area of the circle of radius 1. What is the area of the circle of radius 2?

STUDENT. $4A$.

SERGE LANG. Correct! What's you name?

STUDENT. Ho.

SERGE LANG. Ho. Good for you. $4A$. And what is the area of the circle of radius 10?

ANOTHER STUDENT. $100A$.

SERGE LANG. And what is the area of a circle of radius 1/2? Rachel.

RACHEL. One quarter A.

SERGE LANG. Right. And what is the area of the circle of radius r?

RACHEL. r^2?

SERGE LANG. r^2? [*Questioning*]

RACHEL. r^2A.

SERGE LANG. r^2A. That's it. Now do you see the r^2 coming in? That's the r^2.

By the way, there is a confusion of course about what we mean by a circle. The periphery of the circle, that's what is called the circle. If you want the inside of the circle, we have to make a distinction, we call it a disc. Like a frisbee. We call it a disc because the Greeks called it a disc. If the Greeks had called it a frisbee we would have called it a frisbee.

So we can say that the area of a disc of radius r is r^2A, where A is the area of a disc of radius 1. Well, you want to give a name for the area of the disc of radius 1, and that's what we are going to call pi. So I define π to be the area of a disc of radius 1. For our purposes today, we are going to assume π as being that number. Of course we assume that we have given units of measurement, whichever way you want. And that is what we will take as the definition of π. The area of the disc of radius 1.

Therefore, if we do that, then by what we have seen before, the area of the disc of radius r is πr^2. Before we called A the area of the disc of radius 1. Now we call it π. That's the constant. And we have explained the r^2, starting with this definition of π. This takes care of one of our formulas. OK? Any questions? Can I go on? [*No questions.*]

So let's go back to our original problem, when two questions were raised. The area of a disc, and also the circumference of the circle. We have taken care of the area. There remains the circumference. And the *theorem* is that:

Theorem. *The circumference of a circle of radius r* is $2\pi r$.

That is now the theorem. Let's call the circumference c. We started with a definition of π, we got the πr^2, and we now want to prove that

$$c = 2\pi r.$$

That's what I want to do: give a systematic treatment of the formulas πr^2 and $2\pi r$. Right? Any complaints?

[*Laughter.*]

STUDENT. No complaints.

ANOTHER STUDENT. Not yet!

SERGE. Why do you have the radius, not the diameter?

SERGE LANG. I could have written the formula as $c = \pi d$, where d is the diameter.

STUDENT. But why the diameter? How do you know it is $2r$?

SERGE LANG. You don't. That's exactly what I am going to prove! I am making an assertion; I haven't proved it yet. That is the whole point of the operation. To prove that this is the correct formula. I haven't proved it yet. The proof is going to take place. I don't know, it's quite a remarkable fact, that the same constant that appeared in the formula for the area is going to come in the formula for the circumference, exactly in that pattern. That is precisely the content of the theorem. That it comes in the formula in exactly this manner: the circumference is πd, pi times the diameter.

At first there is no reason why it should come out like that. I mean, it could have come out any other way. If it comes out like that, don't blame me. I mean, God made it that way! [*Laughter.*] I didn't fix it up that way. It was fixed up by somebody else. But you ask me why, and I'm going to answer why. That's what the word "proof" means. [*Laughter.*] So I write the word:

Proof.

And I'll give the proof. What's your name?

STUDENT. Sheryl.

SERGE LANG. So Sheryl asks why. [*Strong laughter.*] Serge also asks why. [*Laughter.*] And I am going to answer why. So what method am I going to use? I'll use the approximation method again. The only thing we are sure of about areas is the area of rectangles and triangles. That's the thing we are going to work with. So I look at my circle of radius r, and I approximate it with triangles. At first, let's use four triangles, like this.

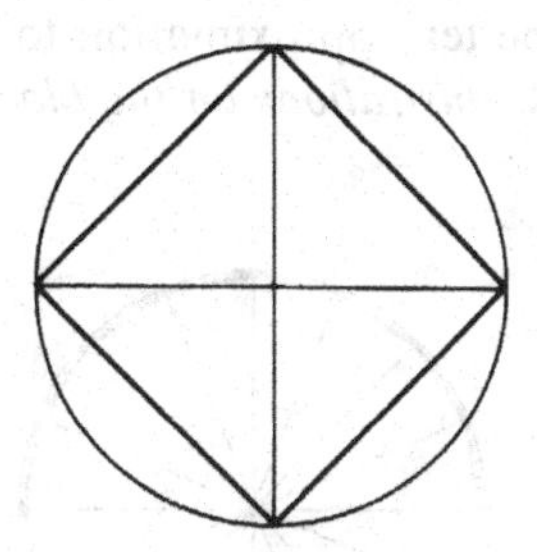

You see four triangles. It's a very rough approximation. It's not especially good. Then I can take my circle of radius r and approximate it with more triangles, like for instance six triangles.

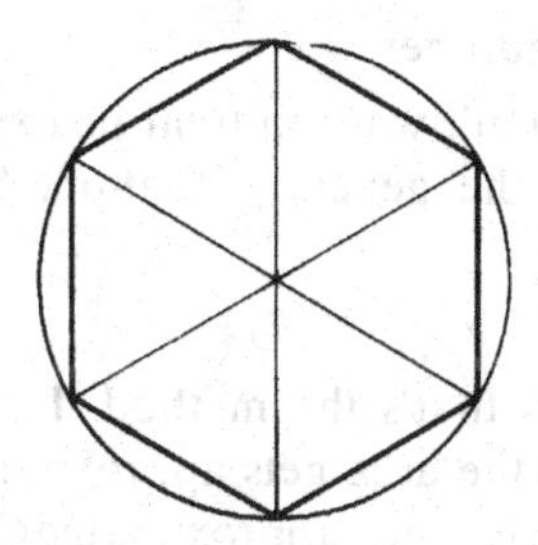

I can take an approximation with seven, or eight

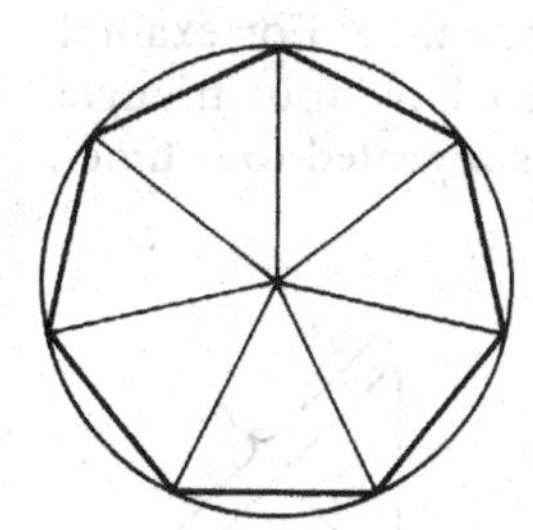

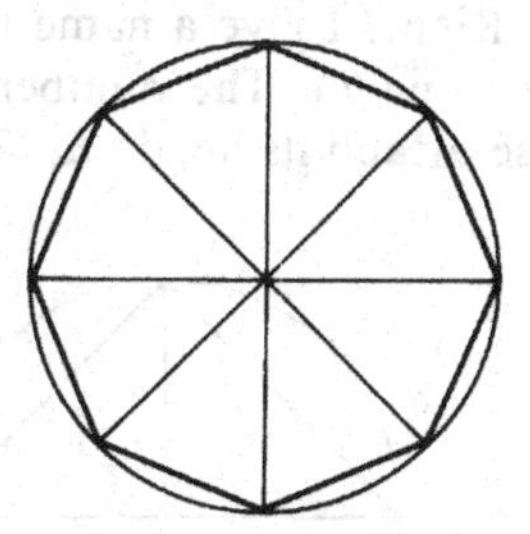

or more like these, which get better and better.

Have you seen regular polygons before? Yes? Good. So I approximate the circle with polygons. The first polygon I used was a square, with four sides. If the regular polygon has six sides, what do you call it? Do you know what that is called, a regular polygon with six sides?

A STUDENT. Hexagon.

SERGE LANG. That's right. A hexagon. I'm just checking what you know. So here are the four triangles, next are the six triangles, seven triangles, and of course I can take more. As I take polygons with more and more sides, I get a better and better approximation to the circle. Right? [*During all this, Serge Lang draws illustrations on the blackboard.*] Here I draw one with n, with arbitrary n.

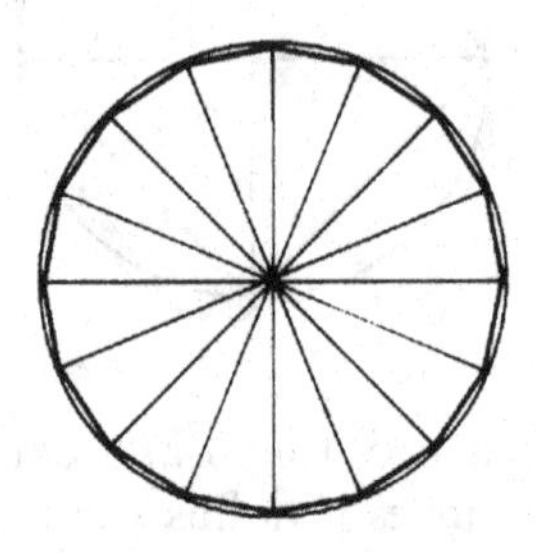

You know about n?

A STUDENT. It's any number.

SERGE LANG. Yes. I don't want to treat just specific cases, I want to be able to take account of the general situation. So I will use n. Does it bother you?

A STUDENT. No.

SERGE LANG. OK. So that's the method. I approximate the circle by regular polygons. And so the area gets approximated by triangles, and the circumference of the circle gets approximated by the perimeter of the polygon. Here with the square it's not very good, we have a certain triangle which is repeated four times. In the next picture, we have a certain triangle which is repeated six times. Over there the triangle is repeated seven times. Right? I give a name to the triangle each time. For example, I call the first one T_4. The number 4 indicates that I have four triangles. I call its base b_4 and its height h_4. The triangle T_4 is repeated four times.

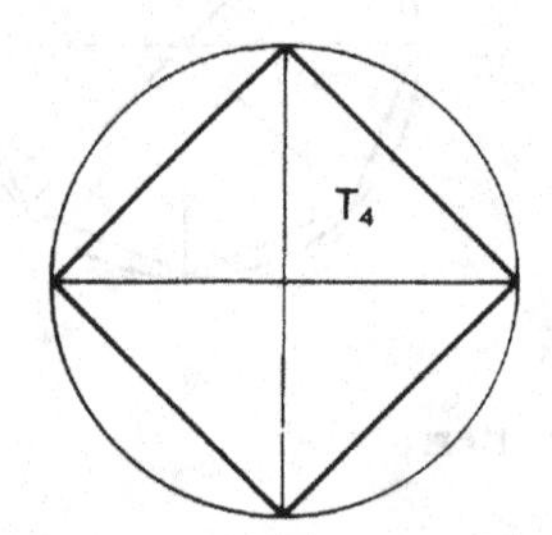

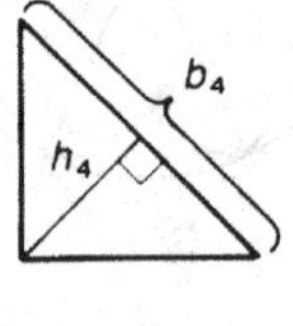

Then I have the triangle T_6, with base b_6 and height h_6. The triangle T_6 is repeated six times.

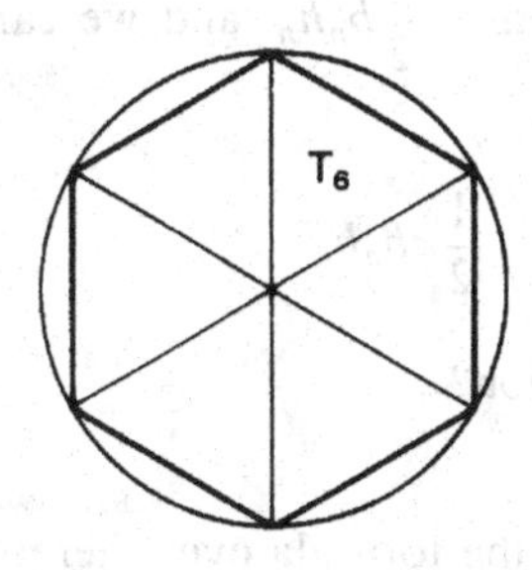

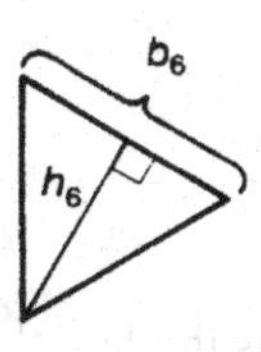

Over there I have the triangle T_7 which is repeated seven times. In general I have the triangle T_n, which is repeated n times. I will call its base b_n and its height h_n.

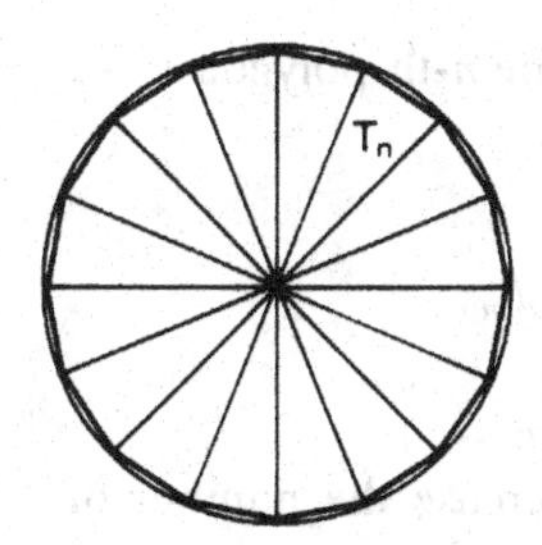

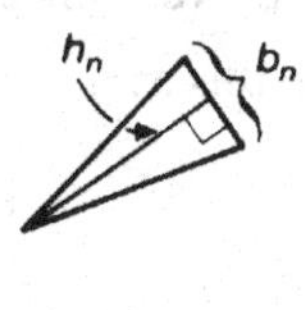

What is the area of T_4 ? What's the area of a triangle?

SHERYL. One half b_n times h_n.

SERGE LANG. You are way ahead of me. Let's go slowly for everybody. The area of T_4 is $\frac{1}{2}b_4h_4$. She gave me the general answer. Let's go step by step. What's the area of T_6 ?

A STUDENT. One half b_6 times h_6.

SERGE LANG. Yes, and what's the area of T_7 ? It's $\frac{1}{2}b_7h_7$. And finally what's the area of T_n ? Sheryl already gave it to me.

A STUDENT. One half h_n times b_n.

SERGE LANG. That's right. Now can I leave the picture just with n, and erase the others, because I need the blackboard space? I picked the pictures with 4, 6, 7 just to lead to the n, but I only need the n. Right?

Now what is the area inside the polygon?

STUDENT. The area of T_n times n.

SERGE LANG. That's right. What's your name?

STUDENT. Charlie.

SERGE LANG. So Charlie said, rightly, that the area inside the polygon is n times the area of T_n. Which is n times $\frac{1}{2}b_nh_n$, and we can write the formula

$$A_n = n\frac{1}{2}b_nh_n = \frac{1}{2}nb_nh_n.$$

And what is the length of the n-th polygon?

SHERYL. n times b_n.

SERGE LANG. That's right. I'll write the formula over there:

$$L_n = nb_n.$$

Do you follow? Adolph, do you follow?

ADOLPH. Yes.

SERGE LANG. Since the area inside the n-th polygon is $A_n = \frac{1}{2}nb_nh_n$, I can write this as

$$A_n = \frac{1}{2}L_nh_n,$$

where L_n is the length of the n-th polygon.

Suppose I make n very large. So I increase the number of sides of the polygon, and I get a better approximation of the circle.

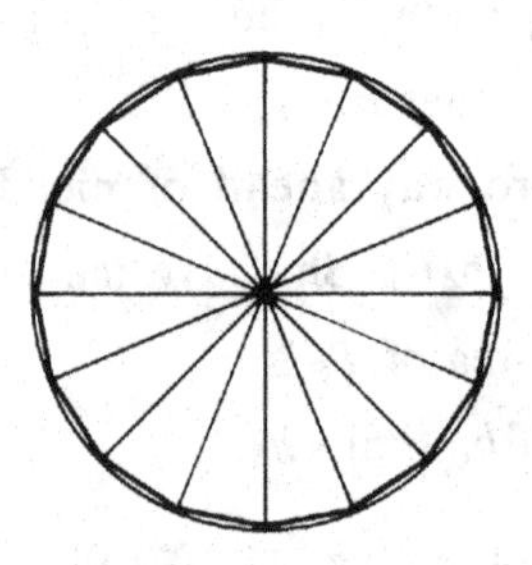

After a while, I can't draw it any more. When n becomes very large, what does this quantity on the right, $\frac{1}{2}L_nh_n$ approach? The length L_n of the polygon approaches what? Charlie.

CHARLIE. The circumference of the circle.

SERGE LANG. Yes. And what does the height h_n approach?

STUDENT. The radius.

SERGE LANG. Very good. What's your name?

STUDENT. Joe.

SERGE LANG. So Joe says that h_n approaches the radius of the circle when n becomes very large. You've been doing very well–Charlie, Joe, Sheryl, Serge... [*Laughter.*], Rachel... It's hard to get all the names straight, for the first time. So h_n approaches the radius [*Serge Lang writes on the blackboard and runs out of chalk.*]

A STUDENT. There's more chalk in the drawer. [*Laughter.*]

SERGE LANG. Good. So on the right hand side of the equality

$$A_n = \frac{1}{2}L_n h_n,$$

the length L_n approaches the circumference c, and h_n approaches the radius. On the left hand side the area A_n inside the polygon approaches–what? Let's pick on somebody new. [*Serge Lang points to another student.*]

THE STUDENT. Pi? [*Serge Lang raises his eyebrows.*] No...

SERGE LANG. I have a circle of radius r. We've decided that the area of the circle of radius r is?

THE STUDENT. πr^2.

SERGE LANG. Good. So the left hand side approaches πr^2, and the right hand side approaches $\frac{1}{2}cr$. Therefore I get

$$\pi r^2 = \frac{1}{2}cr.$$

Now what do you do by algebra? Charlie.

CHARLIE. You cancel the r's.

SERGE LANG. Yes, you can cancel one r on each side, on r on the left and the r on the right, so you get

$$\pi r = \frac{1}{2}c.$$

Then what do you do?

CHARLIE. Multiply by 2.

SERGE LANG. Yes, you multiply by 2, and what do you get?

CHARLIE. $2\pi r$ equals c.

SERGE LANG. Yes, $2\pi r = c$, and that's the formula you wanted to prove. [*Laughter throughout the class.*] You see it?

We started with πr^2 as the area of the disc of radius r. Then we used an approximation, by computing the perimeter of the polygon, the area of the polygon. Then we let n tend to infinity and get the approximation of the length of the circumference and the area inside the circle. The right hand side approaches $\frac{1}{2}cr$; the left hand side approaches the area of the disc which is πr^2. We have already proved that before. And then you use a little algebra, you cancel one r and multiply by 2, and you get the formula

$$2\pi r = c.$$

There is your formula. Do you agree that's a proof? [*Serge Lang points to Rachel.*]

RACHEL. Yes. [*Her tone is uncertain.*]

SERGE LANG. You do?

RACHEL. Yes. [*Laughing a little.*]

SERGE LANG. What do you mean "yes"? Is it a yes by intimidation or a yes by conviction? Or a little bit of both?

RACHEL. A little bit of both. [*Laughter.*]

SERGE LANG. Well, where is the intimidation?

RACHEL. I don't know.

SERGE LANG. You don't know? [*Laughter.*] All right, let's make it all conviction. Look, where did I start from? You granted me that πr^2 is the area of the circle of radius r. Now after that what did I do? Heh?

RACHEL. You drew a circle. [*Laughter.*]

SERGE LANG. Yes, I drew a circle. And after that what did I do?

RACHEL. You divided it into triangles.

SERGE LANG. That's right, I divided the inside of the circle into triangles. I drew a regular polygon inside the circle, which I divided into triangles.

RACHEL. Yeah.

SERGE LANG. Then, I used logic. I mean, you granted me the area of triangles. So I took a regular polygon with n sides, 4, 5, 6, 7, 8, 9, 10... and then in general, I do it with n. "n" doesn't bother you?

RACHEL. No.

SERGE LANG. Not at all? OK. Then I give names to the height and the base of the triangles. I call the base b_n and I call the height h_n. Does that bother you?

RACHEL. No.

SERGE LANG. Is that intimidation or conviction?

RACHEL. That's conviction.

SERGE LANG. So far no intimidation?

RACHEL. So far so good.

SERGE LANG. So far so good. [*Laughter.*] Now what's the area of each triangle?

RACHEL. Half the base times the height.

SERGE LANG. Half the base times the height, you said it yourself. So that's not intimidation.

RACHEL. No, that's conviction.

SERGE LANG. I'm glad that's conviction. [*Laughter.*] All right, what's the area inside the polygon? How many triangles do I have inside? [*Serge Lang draws a new picture.*]

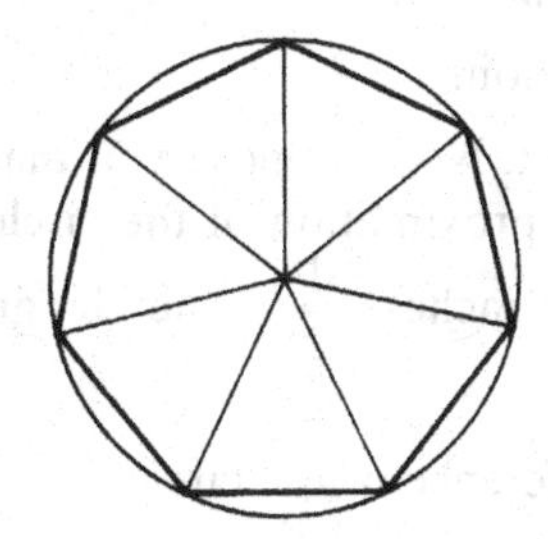

RACHEL. [*She starts counting:*] One, two, three, . . . ,seven–I mean n.

SERGE LANG. n triangles, right. And if I have the area of each triangle, and find n triangles with this same area, then what is the total area inside the polygon?

RACHEL. n times the area of each triangle.

SERGE LANG. So the area inside the polygon is n times $\frac{1}{2}b_n h_n$. So far so good?

RACHEL. Yes.

SERGE LANG. Conviction?

RACHEL. Yeah [*smiling*].

SERGE LANG. All right. Well, what was my next step? The $1/2$ is $1/2$, and then I took n times b_n, n times that base. What is the length of the perimeter of the polygon? Do you see the polygon? How many sides does it have?

RACHEL. Seven–oh, I'm sorry, I mean n.

SERGE LANG. That's all right, you start with the special number 7, but then your mind works it out in general and you say n, which shows you understand. You end up saying there are n sides . . .

RACHEL. Yes.

SERGE LANG. Each side has length b_n. So what is the length of the perimeter of the polygon?

RACHEL. n times b_n.

SERGE LANG. Yes, do you see the n times b_n here?

RACHEL. Yes.

SERGE LANG. So I get nb_n, which is L_n, the length of the n-th polygon, that's the perimeter of the n-th polygon. Then the area inside the n-th polygon is

$$A_n = \frac{1}{2}L_n h_n.$$

Is that conviction or intimidation?

RACHEL. It's a conviction.

SERGE LANG. All right. Now suppose you make n larger and larger to get a better and better approximation of the circle. What do these quantities approach? $\frac{1}{2}$ approaches $\frac{1}{2}$. The length L_n of the polygon approaches what?

RACHEL. The circumference of the circle.

SERGE LANG. [*To the class*:] That's what she said, the circumference of the circle. [*To Rachel*:] Is that conviction or intimidation?

RACHEL. Conviction.

SERGE LANG. The h_n approaches what?

RACHEL. The radius.

SERGE LANG. Definitely. So the product of these three things approaches one half times the circumference times the radius. Do you see it? Any questions so far?

RACHEL. No.

SERGE LANG. Is that conviction or intimidation?

RACHEL. Conviction. [*Smiling*.]

SERGE LANG. All right. Now I have the area A_n inside the n-th polygon on the left hand side, which approaches what?

RACHEL. The area of the circle.

SERGE LANG. Which is what?

RACHEL. πr^2.

SERGE LANG. OK. So I get that πr^2 is equal to $\frac{1}{2}cr$.

RACHEL. So far so good!

SERGE LANG. So far so good. Now if I have

$$\pi r^2 = \frac{1}{2}cr,$$

you should have an irresistible impulse to do something to this equation. [*Laughter.*] What's your irresistible impulse?

RACHEL. To change this to $2\pi r = c$.

SERGE LANG. Yes, and how does this change come about?

RACHEL. You cancel the r, and then you multiply by 2 on one side and by 2 on the other side, and then $2\pi r = c$.

SERGE LANG. That's the formula I wanted to prove!

RACHEL. You proved it. [*Laughter.*]

SERGE LANG. I proved it. Now where is the intimidation?

RACHEL. It's gone. [*Said with definiteness.*]

SERGE LANG. It's gone!? Ah... [*Laughter.*] So we won, the mathematics won! Right?

RACHEL. Right.

[*Serge raises his hand.*]

SERGE LANG. Yes?

SERGE. That's only approximate.

SERGE LANG. No. At the n-th level it's only approximate, but if I take the limit, what the quantities approach, it is not approximate.

SERGE. But still, my point is that it will never quite reach the circle. If you divide the circle by a certain number... even an infinite number...

SERGE LANG. I don't divide by an infinite number. I use the word "approach". Do you grant me that the area inside the polygon approaches the area of the disc?

SERGE. Yes, but it never quite reaches it.

SERGE LANG. No it doesn't, but it approaches it. Whatever the area of the disc is, it is approached by the area inside the n-th polygon.

SERGE. Yes, but it never quite reaches, it's not exact.

SERGE LANG. What is not exact? [*Serge Lang and Serge talk simultaneously.*] I'm going to write something down on the blackboard. I have the area of the circle, which I know is πr^2. I have the area A_n of the n-th polygon. The area A_n approaches the area of the circle. I did not say it is exactly equal, it approaches.

SERGE. Since it never quite reaches, how can you say...

SERGE LANG. I don't understand that sentence. What do you mean "never quite reaches"? [*Serge Lang and Serge talk at the same time.*]

SERGE. The area of the polygon never quite becomes the same as the area of the circle.

SERGE LANG. No, it does not. That's right.

SERGE. OK, then how can you say that your formula is exact?

SERGE LANG. Ah, the area of the n-th polygon approaches the area of the circle. The area of the n-th polygon is $\frac{1}{2}L_n h_n$. What does $\frac{1}{2}L_n h_n$ approach?

SERGE. Well, the, euh . . .

SERGE LANG. What number does $\frac{1}{2}L_n h_n$ approach?

SERGE. $\frac{1}{2}cr$.

SERGE LANG. That's right. On the one hand the area of the n-th polygon approaches πr^2; and on the other hand this other number $\frac{1}{2}L_n h_n$, which is equal to that area, approaches $\frac{1}{2}cr$. If I have an expression which approaches two numbers, but the same expression, then the two numbers have to be equal.

SERGE. Oh, I see what you mean.

SERGE LANG. When I write

$$A_n = \frac{1}{2}L_n h_n,$$

I have an equality sign here. This equality sign is not approximate. This is a precise equality. Yeah? And on the one hand, this number A_n approaches πr^2; and on the other hand that same number $\frac{1}{2}L_n h_n$ approaches $\frac{1}{2}cr$. So the same number approaches two possibilities. Then the two possibilities have to be equal.

SERGE. Ah, yes.

SERGE LANG. Do you see that? Does that answer your question?

SERGE. Yes, it does.

SERGE LANG. Anybody else have a similar question? Do you understand the argument? [*Serge Lang points to a student.*]

THE STUDENT. Yes.

SERGE LANG. What's your name?

STUDENT. Mike.

SERGE LANG. Will you repeat the argument? Mike! [*Laughter.*]

MIKE. OK. You've got a circle, and [*Laughter*] and you divide it, OK, and the radius is r. And you divide it up n times. Here n is any number. So, then with those little... when you divide it you have little triangles, and the area of the triangle is half base times height.

SERGE LANG. Yes.

MIKE. And the base of that triangle would be b_n, the height would be h_n, so the area is half $b_n h_n$.

SERGE LANG. Right.

MIKE. Then you times it by n, which is the number that you got. [*Laughter.*] And then n times b_n is the length, so that becomes L_n. Then half $L_n h_n$ is half times c times r...

SERGE LANG. Approaches, not "is".

MIKE. Oh, approaches. Half times c times r. And the, OK, πr^2 equals $\frac{1}{2}cr$. You cancel one r from each side; then you times both sides by 2, so $2\pi r$ equals c.

SERGE LANG. Bravo! You really got it! [*To the class:*] Do you see what Mike did? He was able to repeat the whole proof, all at once, the whole sequence of ideas. That's wonderful. Good for you.

Comments

The response to the lecture was especially gratifying. A teacher told me afterward that she had been apprehensive about Rachel, when I started the "conviction–intimidation" sequence, whether Rachel would bear up under the strain. Not only did Rachel bear up (I don't even know if it was a strain), but she progressively acquired confidence, and the tone of her answers changed from uncertain to being quite assertive. Regretfully, the tone cannot be reproduced on the printed page.

I would not engage in this kind of sequence with any student. There has to be a quick judgment on my part whether the student will indeed find such an exchange too much of a strain, or on the contrary, will reach a new level of understanding because of it.

The final performance by Mike, who is able to reproduce the whole proof, is a great success. One of the objections which I have to elementary and secondary school curriculum and teaching is the overemphasis on technical matters, the incoherence, and the lack of sweep (analogous to musical phrases) in the questions considered. The topic I selected has a certain sweep, and it was fascinating to see the mental wheels turning, reflected in Mike's facial expression, while he was reproducing the proof. He had to do it all verbally, put the sentences together, and solve a number of problems of organization, choice of words, sequence of ideas, which gave pleasure to him, to the rest of the class, and to me.

On a more technical level, the approximation discussed in the first part of the proof (concerning the change of area under dilation) should be pushed a little beyond the point where I did, but I did not have the time to do so since I wanted to deal completely with the second part of the proof. If I had another hour at my disposal, I would expand on the reasons why the area so changes along the following lines. Consider a grid consisting of vertical and horizontal lines:

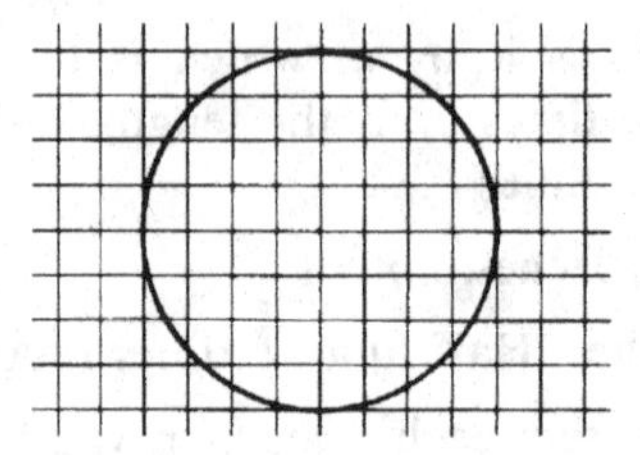

To determine the area of the disc approximately, we could count all the squares that lie inside the circle, measure their sides, add up their areas, and get the desired approximation. But we want to estimate how good is the approximation. The difference between the sum of the areas of all the little squares contained in the disc and the area of the disc itself is determined by all the small portions of squares which touch the boundary of the disc, i.e. which touch the circle. We have a very strong intuition that the sum of such little squares is quite small if the grid is fine enough. We can give an estimate for this smallness. Suppose that we make the grid so that the squares have sides of length a. Then the diagonal of such a square has length $a\sqrt{2}$. If a square intersects the circle, then any point on the square is at distance at most $a\sqrt{2}$ from the circle. Look at figure (a).

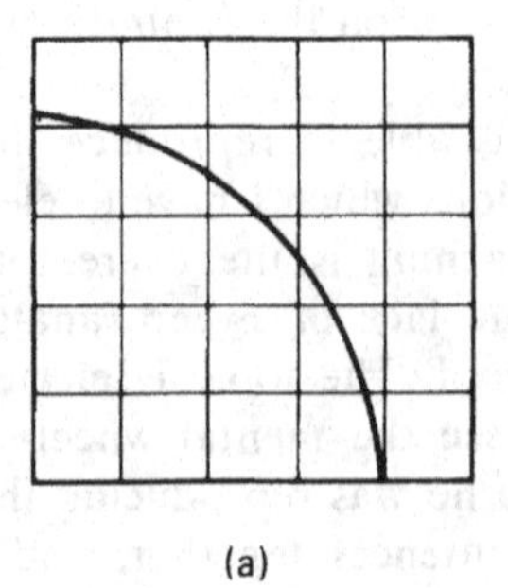

(a)

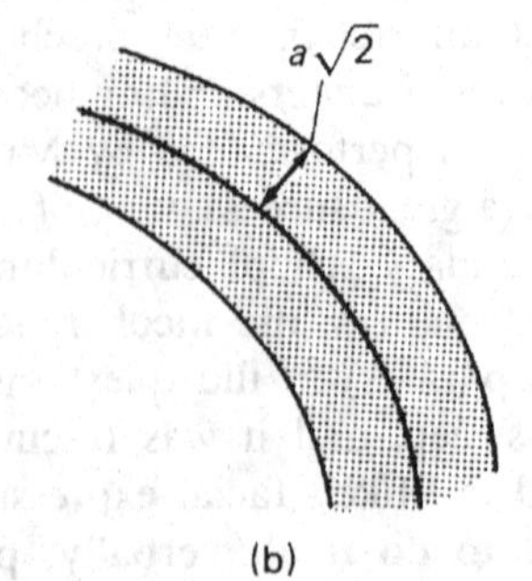

(b)

This is because the distance between any two points of the square is at most $a\sqrt{2}$. Let us draw a band of width $a\sqrt{2}$ on each side of the circle, as shown in figure (b). Then all the squares which intersect the circle must lie within that band. It is very plausible that the area of the band is at most equal to

$$2a\sqrt{2} \text{ times the length of the circle.}$$

Thus if we take a to be very small, i.e. if we take the grid to be a very fine grid, then the area of the band is small, namely at most $2a\sqrt{2}L$ where L is the circumference of the circle.

Therefore we have estimated the error in approximating the area of the circle (disc) by the sum of the areas of the squares of the grid which intersect the circle. This error approaches 0 as the size of the grid approaches 0. Therefore, as the size of the grid approaches 0, the sum of the areas of the squares of the grid which lie entirely inside the circle approaches the area of the disc.

A similar argument applies to any curved figure.

Finally, in a later talk, you will see another way of deducing the length of the circle $c = 2\pi r$ from the area of the disc $A = \pi r^2$, which in some ways is simpler than the one chosen here, and which generalizes to finding the area of the sphere.

Volumes in higher dimension

This is the second talk, following "What is pi?" and dealing with volumes of standard figures like pyramids and cones. The talk was given to the same class, with 10th grade students, approximately 15 years old, in a high school in the suburbs of Toronto.

SERGE LANG. Today, we are going to do similar things, but in higher dimension. But first, a remark in 2 dimensions. Let's start again with our rectangle, with sides a, b.

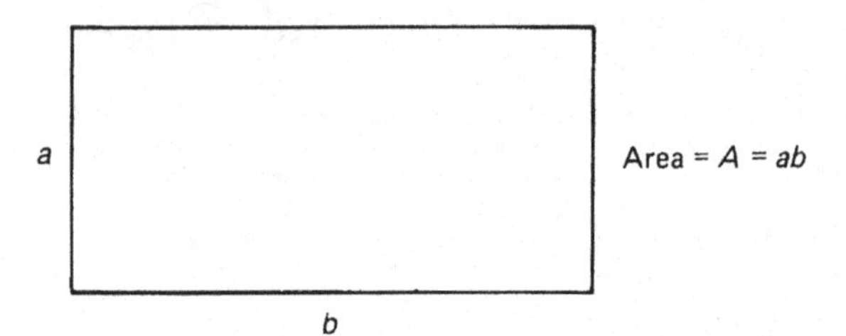

Yesterday, I made dilations by a factor of r in all directions. It could be a blow up or a shrink down–I called it a dilation. Now you should note that I can dilate one direction by a certain factor, and another direction by another factor. For instance, I can keep one direction fixed, and blow up in the other direction by a factor of 2. Then this side has length $2a$, but the other side still has length b.

If the area of the first rectangle is A, what is the new area? Sheryl.

SHERYL. $2A$.

SERGE LANG. Yes, $2A$. In general, if I make a blow up by a factor of r in one direction, and s in the other direction, the area will change by what factor?

SHERYL. rsA ?

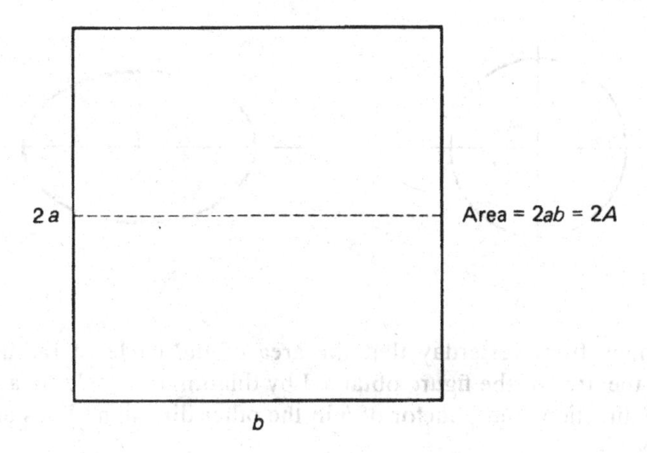

SERGE LANG. That's right. The area of the new rectangle will be rsA.

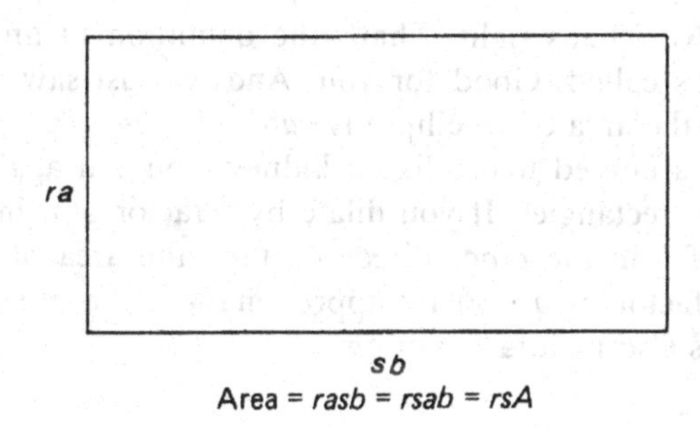

So let's write this down. If we dilate one direction by a factor of r, and if we dilate the other direction by a factor of s, then area changes by a factor of rs.

See, yesterday, we dilated both directions by a factor of r, and we saw that area changes by a factor of r^2. Now today we allow dilations in one direction by a factor of r, and in another direction by a factor of s. Is that clear? By directions, I mean that we have picked two directions in the plane, which are perpendicular to each other.

Like yesterday, we started with rectangles. What happens for general areas? Let's go back to the kidney, or even more simply, suppose I start with a circle of radius 1, and I dilate the circle by a factor of a in one direction, and a factor of b in another direction, like this:

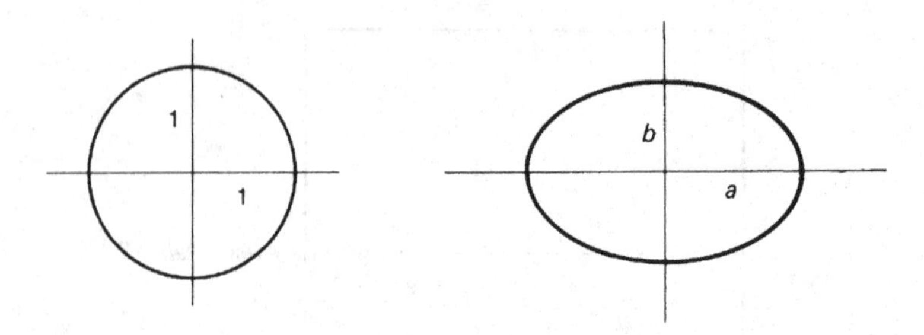

We know from yesterday that the area of the circle of radius 1 is π. What is the area of the figure obtained by dilating the circle by a factor of a in one direction, and a factor of b in the other direction? Let's ask somebody. Joe?

JOE. πab.

SERGE LANG. πab, that's right! By the way, do you know what it's called, a dilated circle, by a factor of a in one direction and a factor of b in the other direction? You know what it's called?

A STUDENT. An ellipse.

SERGE LANG. That's right. That's the definition of an ellipse. That's exactly what it's called. Good for you. And we just saw, right from the definition, that the area of an ellipse is πab.

If you have a curved figure like a kidney, you can again approximate its area by little rectangles. If you dilate by a factor of a in one direction, and a factor of b in the other direction, then the area of each rectangle changes by a factor of ab, so by approximation, the area of the curved figured changes also by a factor of ab.

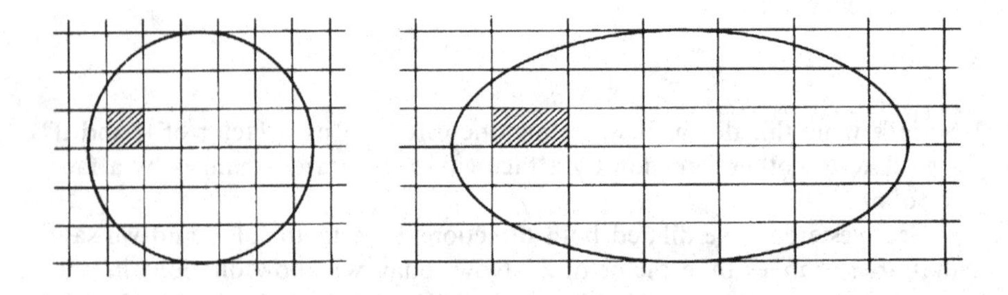

Now let's go to 3 dimensions. We have worked in the plane.

In 3 dimensions, what is the analogous thing to a rectangle? What is the 3-dimensional analogue of a rectangle?

SHERYL. Oh—length times width times depth.

SERGE LANG. Yeah, I have length, width, and depth. So I have 3 directions, and what's the analogue of a rectangle?

[*Some fumbling by several students.*]

SERGE LANG. Well, it's called a rectangular box, OK? Do you have any other name for it?

SHERYL. A cube?

SERGE LANG. Well, a cube is a special case of a box, when all the sides are equal. A cube would be the analogous thing to a square. The analogous thing to a rectangle, I'll just call it a rectangular box. Let's draw one. See the box?

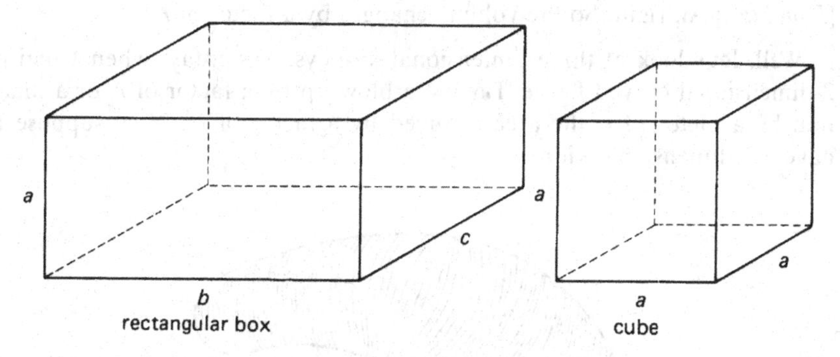

Suppose its sides are a,b,c. Then what is the volume?

SHERYL. abc.

SERGE LANG. That's right, the volume is the product of the sides,

$$V = abc.$$

Now suppose I make a dilation by a factor of r in all directions in three dimensions. So I blow up the whole 3-space by a factor of r.

[*A student starts interrupting getting into volumes. . .*]

SERGE LANG. Wait, wait, wait! Let me draw it. You guys are too fast for me. [*Laughter.*] Then the sides of the dilated box are ra, rb, rc.

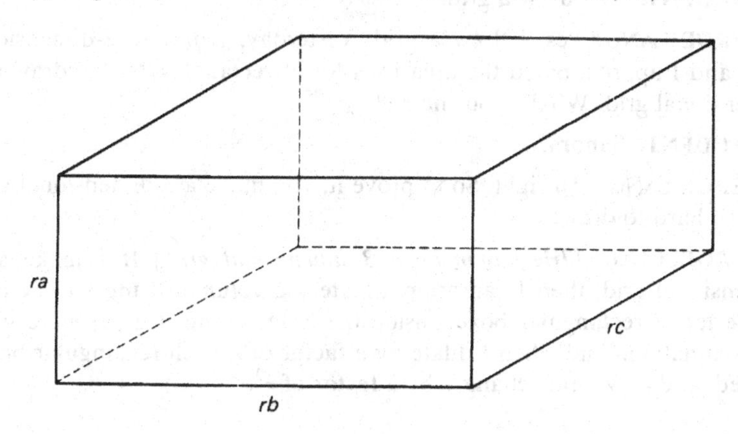

So what is the volume of the box dilated by a factor of r in all directions? Let's pick on somebody. [*Serge Lang points to a student.*]

STUDENT. r^3abc.

SERGE LANG. That's right, r^3abc. Good for you. So the dilated volume is r^3V, if V was the volume of the original box. What's your name?

STUDENT. Gary.

SERGE LANG. Yes, I remember from yesterday. Let's see–Sheryl, Serge, where is Rachel? Oh, over there, you moved! You confused me. [*Laughter.*] All right. So the volume changes by a factor of r^3.

Well, let's look at three dimensional kidneys. Yesterday, when I had a 2-dimensional curved figure, I made a blow up by a factor of r, or a dilation by a factor of r, the area changed by a factor of r^2. Now suppose I have a 3-dimensional kidney.

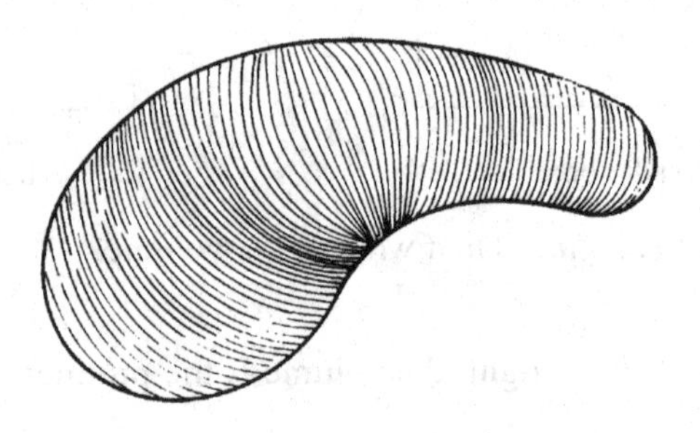

And it has a certain volume V. I make a dilation by a factor of r. What will be the new volume?

SERGE. r^3V.

SERGE LANG. r^3V, again. And how would you prove it? Well how did I prove it yesterday?

STUDENT. You drew a grid.

SERGE LANG. Yes, I drew a grid. Yesterday, I drew a 2-dimensional grid, and I approximated the area by a lot of rectangles. Now I draw a 3-dimensional grid. What's your name?

STUDENT. Sandra.

SERGE LANG. All right. So to prove it, you make a 3-dimensional grid. But it's hard to draw!

SERGE LANG. [*Tries to make a 3-dimensional grid.*] If I make a 3-dimensional grid, then I can approximate the volume of the kidney by a whole lot of rectangular boxes inside the kidney, and it gives me a good approximation. And when I dilate by a factor of r, each rectangular box is dilated, and its volume changes by a factor of r^3.

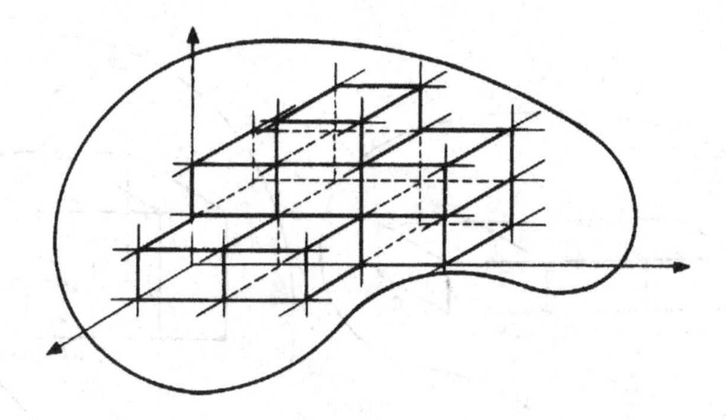

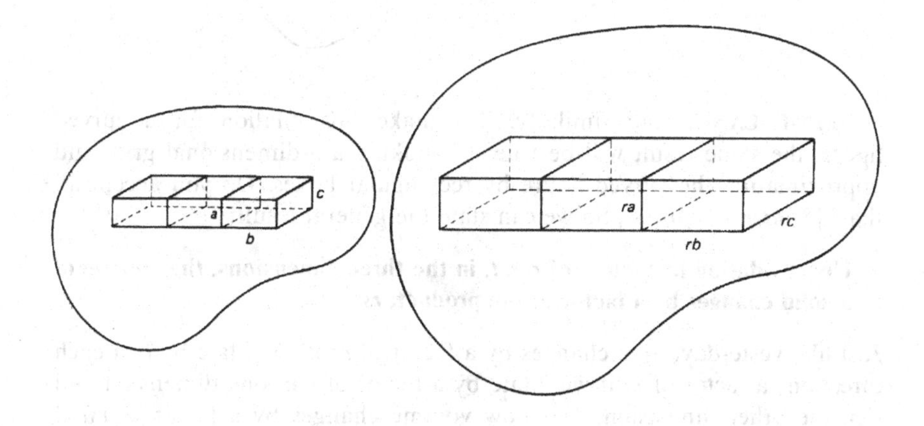

This will be true for each one of the little boxes. And since all the little boxes approximate the volume of the kidney, that means that the actual volume of the kidney changes also by a factor of r^3. So do you accept that? But I have difficulties drawing it.

By the way, just like in two dimensions, I could make a dilation by a factor in one direction, and another factor in the other direction. So let's draw it here.

I start with a rectangular box, and I make a dilation by a factor of r in one direction, s in another, and t in the third direction. If the volume of the original box is V, what is the volume after I make these dilations?

STUDENT. $rstV$.

SERGE LANG. Yes, so the volume changes by the product of the three factors.

ANOTHER STUDENT. Yes, the volume is $rasbtc$, so $rstabc$ and the volume changes by the product rst.

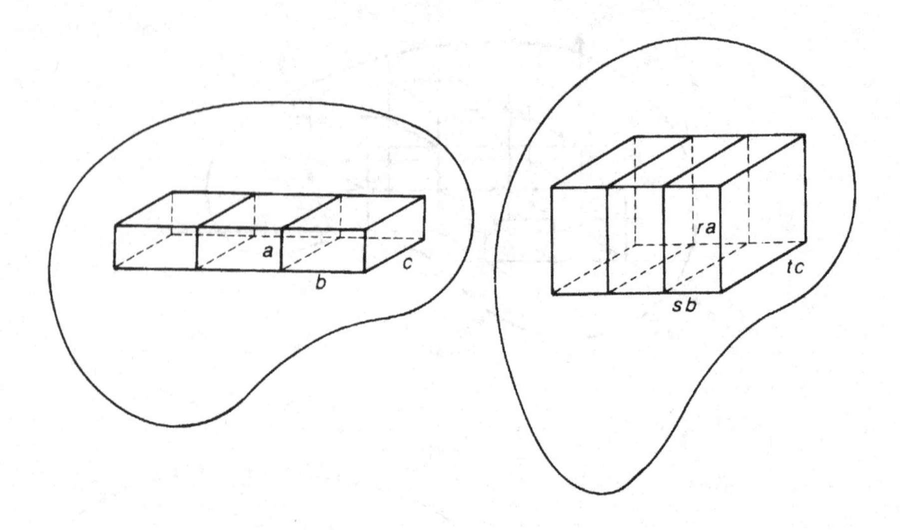

SERGE LANG. And similarly, if I make this dilation for a curved figure, the same result will be true; by making a 3-dimensional grid, and approximating the curved figure by rectangular boxes. Do you accept all that? [*Students approve.*] So we can state the general result:

Under dilation by factors of r, s, t, in the three dimensions, the volume of a solid changes by a factor of the product, rst.

Just like yesterday: area changes by a factor of r^2 if we dilate by r in each direction; a factor of rs if we dilate by a factor of r in one dimension and s in the other dimension; and now volume changes by a factor of rst if you dilate by a factor of r in one dimension, s in another and t in the third. And the three dimensions are in perpendicular directions.

Now I'll deal mostly in three dimensions, but what would be a natural generalization of this? Serge.

SERGE. I don't know.

SERGE LANG. What's a generalization of what I have just done there? I started in 2 dimensions, then I went to 3 dimensions . . .

SERGE. [*Interrupts.*] Four dimensions. OK. It's the next product. I see. It's rst whatever.

SERGE LANG. Ah, rst whatever. That's right. So suppose I have a solid in four dimensions. You see the four dimensions? Now I can't draw it.

STUDENT. Well, you could not draw it either in three dimensions!

SERGE LANG. That's a very good remark. You are absolutely right. So the truth of what I am saying does not depend on my ability to draw the picture! So suppose I have a solid in four dimensions, and I make a dilation by a factor of r in all four dimensions. How does the volume of the solid change? Sheryl.

SHERYL. r to the fourth, times V.

SERGE LANG. r^4V, where V is the 4-dimensional volume. And suppose I have a solid in n dimensions, and I make a dilation by a factor of r in all directions, in all n dimensions. How does the volume change?

SERGE. r to the power n.

SERGE LANG. r^n, and that's how it is in n dimensions. OK? Any problems? Sandra.

SANDRA. No. [*The other students nod, and seem perfectly at ease.*]

SERGE LANG. All right. So we see how volume changes in n dimensions.

Let's go back to three dimensions, and see how we can find the volume of some standard figures. But I think it's remarkable how you react to the possibility of n dimensions. [*Laughter.*] I am slightly taken aback at the way you just went along with it.

All right, let's try to determine volumes for figures like the pyramid and the cone. Let's take the pyramid. First, take the most simple-minded pyramid, when the base is a square, and the pyramid is straight up, like the Egyptian pyramids. See the pyramid there?

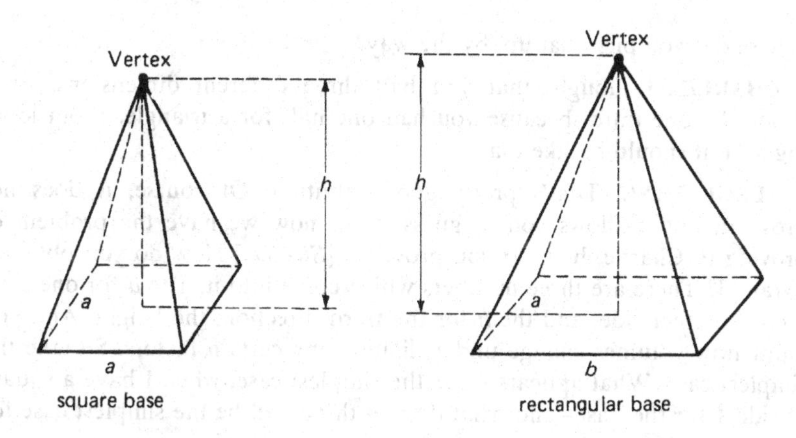

The top of the pyramid is called the vertex. The pyramid has a square base, with sides a, a; and h is the perpendicular height of the pyramid. What do you think is the volume? What's the formula for the volume? Also the pyramid with rectangular base. Ken: what do you guess the volume is?

KEN. One half abh?

SERGE LANG. That's your guess?

KEN. Yes.

SERGE LANG. Sheryl, what do you say?

SHERYL. The same.

SERGE LANG. You say 1/2? What would you say? [*Pointing to a student.*]

CHARLIE. One third.

SERGE LANG. Ah, Adolph says one third. [*Laughter. It was not Adolph* . . .] Euh, sorry, what was your name?

CHARLIE. Charlie.

SERGE LANG. Sorry, I got mixed up. Charlie says 1/3. Any other guesses?

A STUDENT. One quarter *abh*.

SERGE LANG. $\frac{1}{4}abh$?

ANOTHER STUDENT. One fifth *abh*? [*Laughter.*]

SERGE LANG. Well, you could go on quite a while like that. [*Several students discuss the matter among themselves.*] All right, I'll tell you. The answer that is true is the one third. That's what's true:

The volume of the pyramid with rectangular base is $\frac{1}{3}abh$.

Where did you pick that up, by the way?

CHARLIE. I thought that you had three different dimensions, so it should be one third, because you had one half for a triangle. From looking at it, it should be like that.

SERGE LANG. That's pretty good intuition. Of course, it does not prove it, but it allows you to guess it. So now we have the problem of proving it. Charlie, how do you prove it? [*Silence.*] How do you obtain a pyramid? There are three numbers which come into it. The a for one side, b on the other side, and the h for the third direction, the height. And you know how volumes change under dilations by certain factors. So take the simplest case. What appears to be the simplest case, when I have a square of side 1 for the base–and what do you think will be the simplest case for the height? Charlie.

CHARLIE. 3

SERGE LANG. No. Why should the height be 3?

CHARLIE. Because you divide by 3.

SERGE LANG. All right, let's consider that. He wants to make a pyramid of height 3. Is that the simplest thing you can think of? How are we going to prove what the volume is, of this object? See, if you knew the volume of this object, you could get the volume of that one (a dilated one) by dilations.

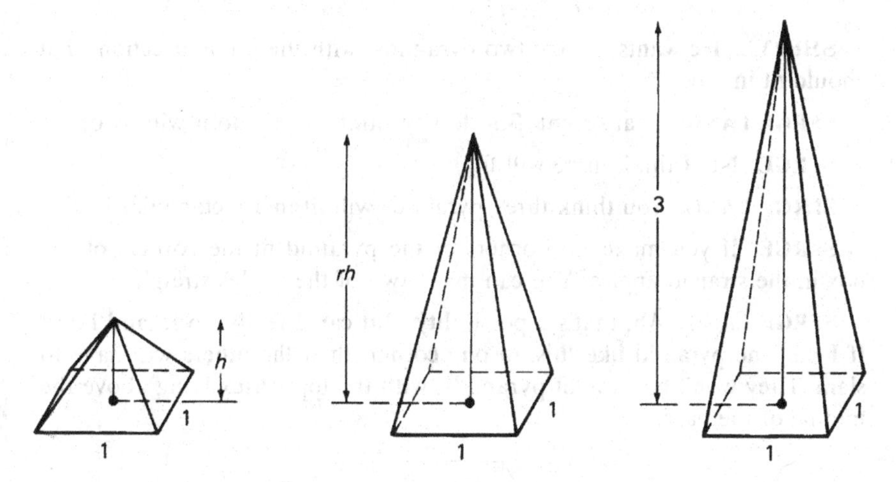

In your pyramid of height 3, there is not enough symmetry. The height is still too arbitrary. You have to try to get as much symmetry as you possibly can find.

SHERYL. The height is 2?

SERGE, OR KEN. One?

SERGE LANG. One looks already better. That's right, that's more symmetric.

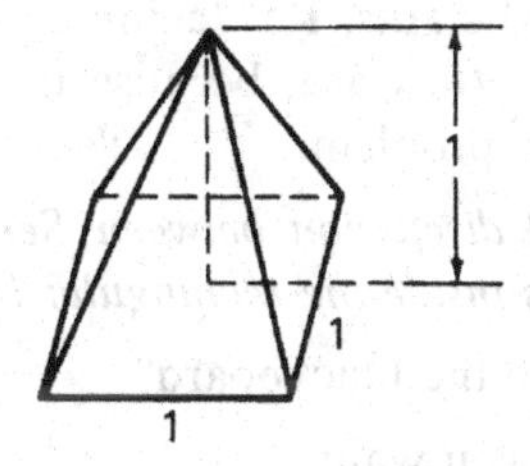

But even with height 1, how are you going to prove it? You still have a problem. So you stare at the problem for a while, and I think eventually you would find how to do it. But I have only forty minutes more, so in fact I'm going to pull a rabbit out of a hat. I mean, I have no choice. If I could meet you once more, I would leave it like this today, tell you to go home and think about it before you go to bed; and if you solve it yourselves, you'd get a kick out of it. Now, I have to interfere with that kick, because I have only forty more minutes.

So the proof is as follows. I take the simplest case [*Serge raises his hand*]—Yes?

SERGE. I think that you could, by taking three pyramids, if you put them together with their corners straight, you'd get a box . . .

SERGE LANG. Eh, eh, eh, wow, you see, he's getting there! What do you want to do? What does Serge want to do?

SHERYL. He wants to take two pyramids, with the same direction, that should fit in a box.

SERGE LANG. That's right. But do you think he can do it with two?

SERGE. No, I think three will fit.

SERGE LANG. You think three pyramids will fit in a rectangular box?

SERGE. If you make the corners of the pyramid fit the corners of the box in the straight angles. You can make two of the angles straight.

SERGE LANG. Ah, that's a possibility. But can I really do it, and how? If I put one pyramid like this, in one corner, then the others will have to slant. They won't be straight pyramids, with the top vertex lying above the middle of the base.

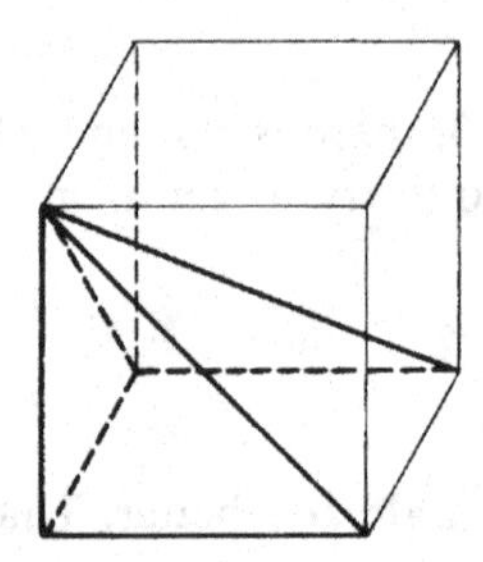

The moment the pyramid slants, I'm facing a new phenomenon, which I will discuss immediately afterward, because it's a nice phenomenon. But the slanting creates a new problem.

[*There follows a rapid discussion between Serge Lang and Serge about how to place the pyramids inside the rectangular box.*]

SERGE. Can I come to the blackboard?

SERGE LANG. Yes, if you want.

[*Serge comes to the blackboard and starts drawing various possibilities. Laughter by the rest of the class at the vivacious exchange. There is an active discussion between Serge, members of the class and Serge Lang about the various possibilities.*]

SERGE LANG. All right, let's come back to what Serge wants to do later. Suppose now we consider this picture.

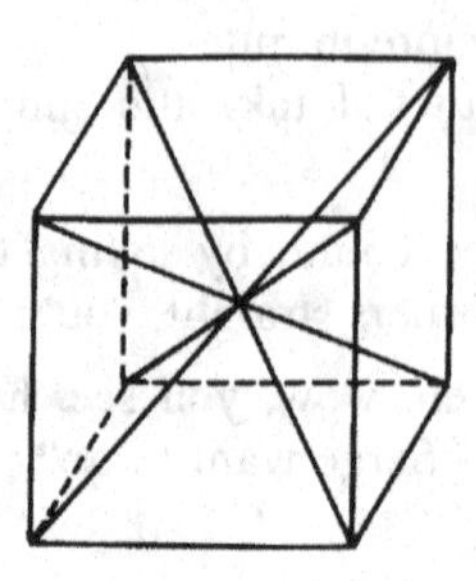

SERGE LANG. Suppose you decompose the box like this, putting the vertex of the pyramid in the middle. How many pieces inside the box will you get?

A STUDENT. You want to know how many parts there are? There are six.

SERGE LANG. Yes. Six together make up the box. What's your name?

STUDENT. Lisa.

SERGE LANG. She's absolutely right. She's going to get it. You have to put six pyramids together to make up a box. Now then, why don't you start in the opposite way, with a box that you know about. What is the simplest box?

LISA. One by one by one.

SERGE LANG. Perfect. So take the simplest box, a cube which is one by one by one.

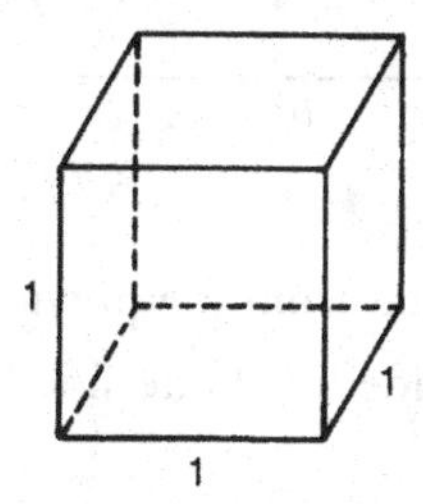

That's what Lisa said. Where will I put the vertex of the pyramid?

LISA. In the middle.

SERGE LANG. Right here, in he middle. Lisa said, put the six pyramids together. I'll do what she said. Like that.

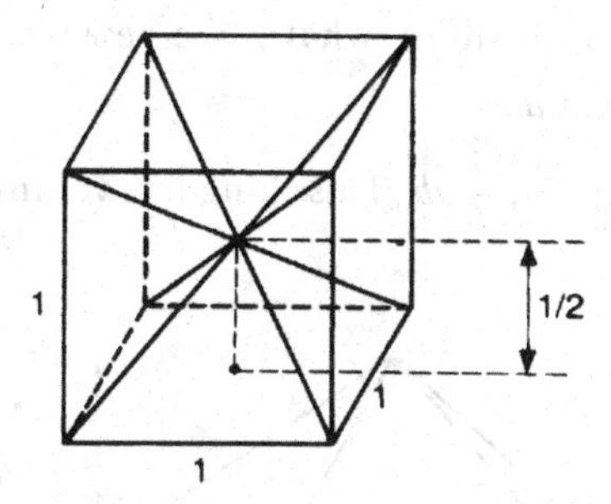

You see, there are six sides to the box, which is a cube. Then there are six pyramids, with vertex at the center of the cube, and whose bases are the six sides of the cube. These six pyramids together fill out the total cube. So what is the volume of each pyramid? Lisa.

LISA. One sixth.

SERGE LANG. 1/6. And the total volume of the whole cube is 1, so each pyramid is one sixth of the total volume, which is 1/6. So now I have the volume of one pyramid, whose base has sides 1, 1—and what is the height of that pyramid?

A STUDENT. One half the side of the cube.

SERGE LANG. Yes, 1/2. The height is 1/2. Now suppose I have a pyramid whose base has sides a, b and the height is still 1/2.

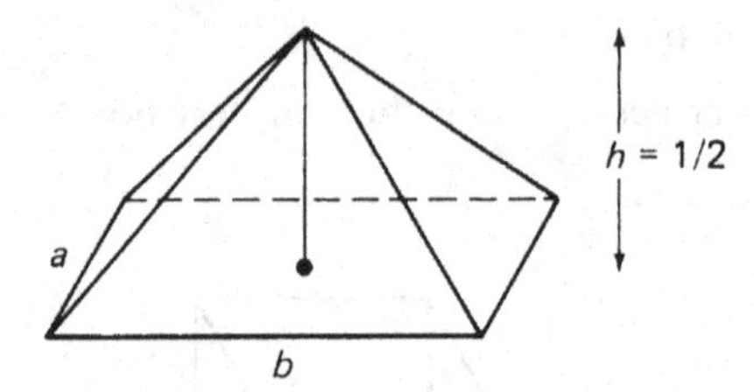

How is this pyramid obtained from the previous one? Mike.

MIKE. By dilating one side by a, one side by b, and the height by nothing.

SERGE LANG. That's right. So the pyramid whose base has sides a, b and with height 1/2 is obtained by making a dilation by a factor of a, b in the two base directions, and a factor of 1 in the third direction. As Mike said. So what is the volume of this pyramid?

MIKE. Half ab?

SERGE LANG. No. The other pyramid, with sides 1, 1 and height 1/2 had volume 1/6. If I make a dilation by a factor of a on one side, b on the other, and 1 on the third, what is the new volume?

MIKE. One sixth ab.

SERGE LANG. Yes, $\frac{1}{6}ab$. This thing has volume $\frac{1}{6}ab$.

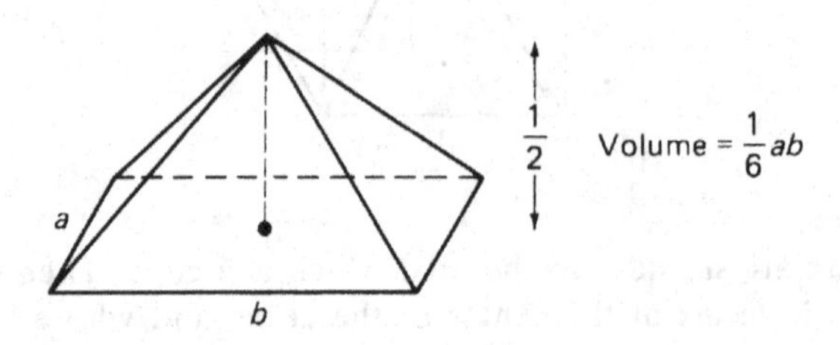

Now suppose I change the height, and I look at the following pyramid, with base ab and the height is 1?

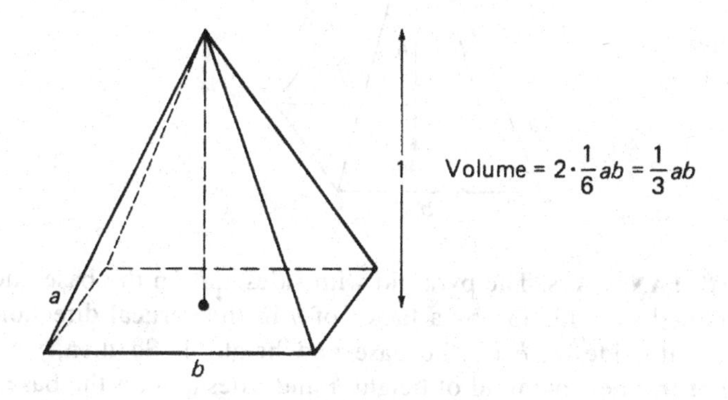

I make a dilation in the third direction, the vertical direction, by what factor?

MIKE. 1.

SERGE LANG. No. I start with the pyramid with height 1/2. The height of the new pyramid is 1. By how much have I dilated in this direction?

MIKE. It's doubled.

SERGE LANG. Certainly, so I have dilated by what factor?

MIKE. 2.

SERGE LANG. That's right. So I have three pyramids. The first with sides 1, 1 and height 1/2. The second with sides a, b and height 1/2. The third with sides a, b and height 1, which is obtained from the second by dilation by a factor of 2 in one direction. The volume of the second pyramid is $\frac{1}{6}ab$. What is the volume of the third pyramid?

MIKE. Two one sixth ab.

SERGE LANG. Yes, $2 \times \frac{1}{6}ab$, which is what?

MIKE. One third ab.

SERGE LANG. $\frac{1}{3}ab$. And you see that one third coming out? We have caught the one third.

Now I draw the final pyramid, with arbitrary height h. How is the pyramid with sides a, b on the base and height h obtained from the pyramid with sides a, b on the base, and height 1? Who knows?

[*Most hands in the class go up.*]

MIKE. Times h.

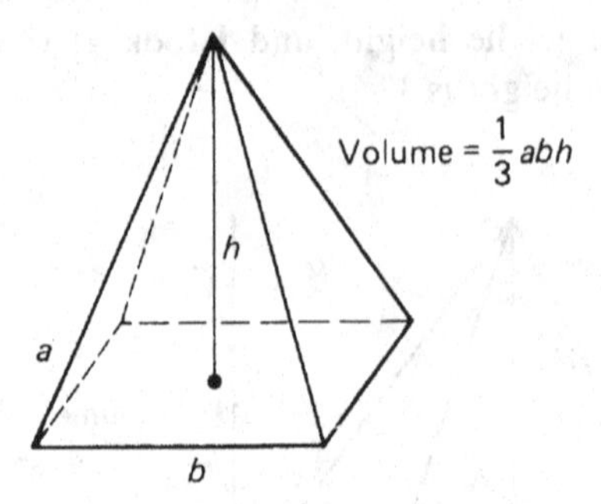

SERGE LANG. Yes. The pyramid with sides a, b on the base and height h is obtained as a dilation by a factor of h in the vertical direction, of the pyramid with sides a, b on the base and height 1. So then, what is the volume of this new pyramid of height h and sides a, b on the base?

SANDRA. One third *abh*.

SERGE LANG. Yes, so we write

$$\textbf{Volume of pyramid} = \frac{1}{3}abh.$$

But that is the most general theorem!

With this method, I take the pyramids with the top at the center of the cube, and I get six pyramids. In this situation, the simplest case was of a pyramid with a base of sides 1 but height 1/2. Then I could get the volume directly, for a pyramid like the Egyptian pyramids, when the top lies directly above the center of the base.

SERGE. I think you could get it a different way. There are three pyramids which make up the cube. The volume of the cube is *abh*.

SERGE LANG. It's not a cube, it's a rectangular box.

SERGE. All right. But you have three pyramids making up the box.

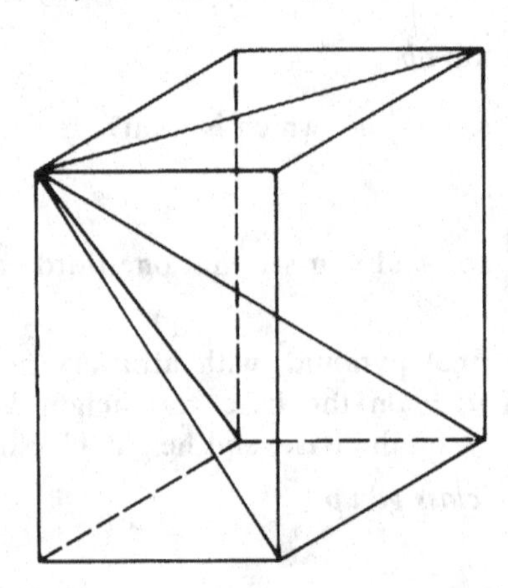

[*Serge draws picture on the blackboard.*] You take the three pyramids with bases which are the sides opposite to one corner of the box.

SERGE LANG. No, not quite, because if you take a, b, and h different, then the pyramids are not all the same. If I take a cube and all its sides are equal to 1, then you do end up with three pyramids which are the same, and your idea is a good one.

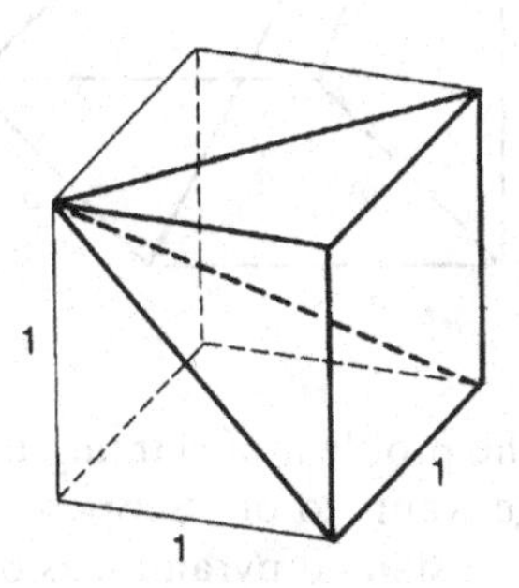

Then by symmetry all three pyramids have the same volume. If you have a rectangular box, not a cube, then you don't have the symmetry. That's why we have to deal with a cube, rather than a rectangular box. To have more symmetry. With your method, the pyramids which you get are slanted, like this:

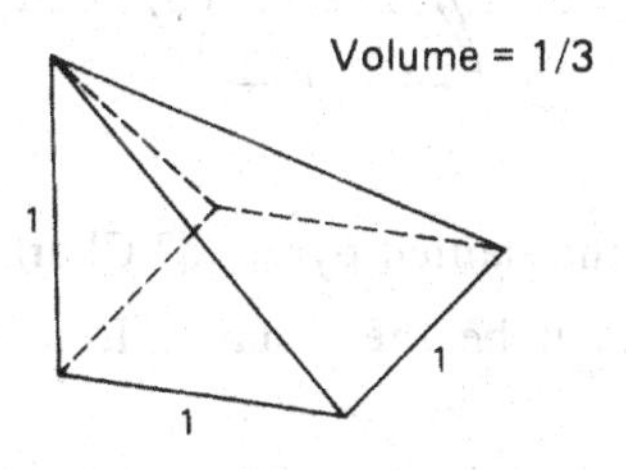

The top of each one of your pyramids does not lie exactly above the center of the base. So you have a different kind of pyramid. But you are right that the volume of each one is one third the volume of the cube, so you get the right answer for pyramids whose top lies vertically above one corner of the base. In this case the pyramid is one third of a cube of side 1.

Then we can make a dilation by a factor of a on one side of the base, a factor of b on another side of the base, and a factor of h vertically, and you get the formula for your type of pyramid, when the vertex lies directly above one corner of the base:

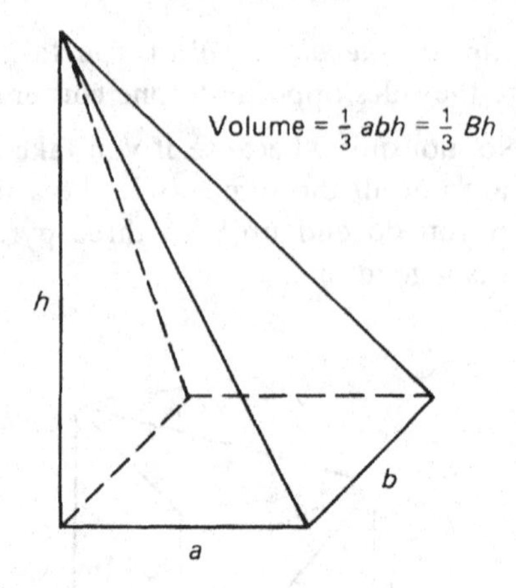

But then you still have the problem of slanting to deal with.

So let's do what Serge wants to do. Suppose I have a rectangle for the base, and suppose I have a slanted pyramid, as on the picture.

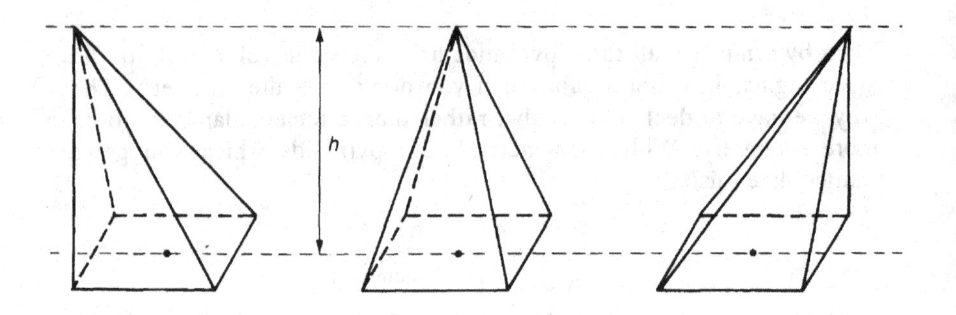

What is the height of the slanted pyramid? Charlie.

CHARLIE. Wouldn't it be the distance from the point on top to the base?

SERGE LANG. That's right. So what will be the volume of the slanted pyramid?

CHARLIE. *abh* one third?

SERGE LANG. Yes, that's the guess, as Charlie said. But we have not proved it. So we have to know about the slanting process, so let's study slanting, which is what Serge wanted to do. It is also called *shearing*.

Let's go back to two dimensions. I take a rectangle.

STUDENT. Not a square?

SERGE LANG. No, not necessarily. You can think of a rectangle as the side of a deck of cards.

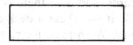

And I slant it. This way.

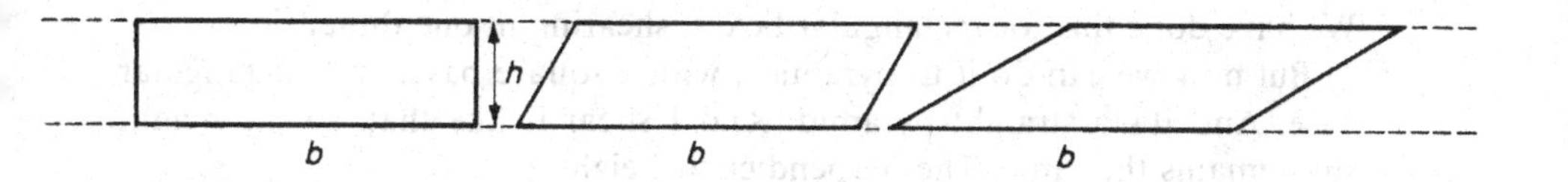

So I get a parallelogram. Then the height is still the same, the perpendicular height. So if A is the area of the rectangle, what's the area of the parallelogram?

SANDRA. A.

SERGE LANG. It's still A. The area does not change. Have you seen that before? You have done the area of a parallelogram, last year?

CHARLIE. Yeah.

SERGE LANG. And how did you prove it? [*Students simultaneously recall the proof with triangles.*] That's right, you use this picture:

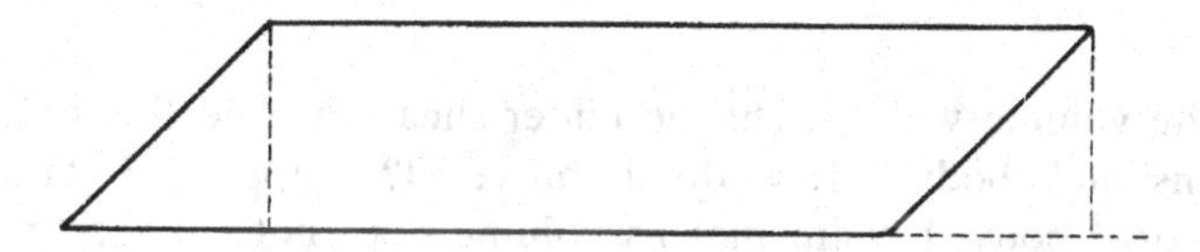

With the two triangles. So we assume that you know that the area of a parallelogram is the base times the height. Then we can state the theorem:

Under slanting, or under shearing, the area does not change.

Now in three dimensions, take a deck of cards, so the rectangle is the side of a deck of cards. And then slant it.

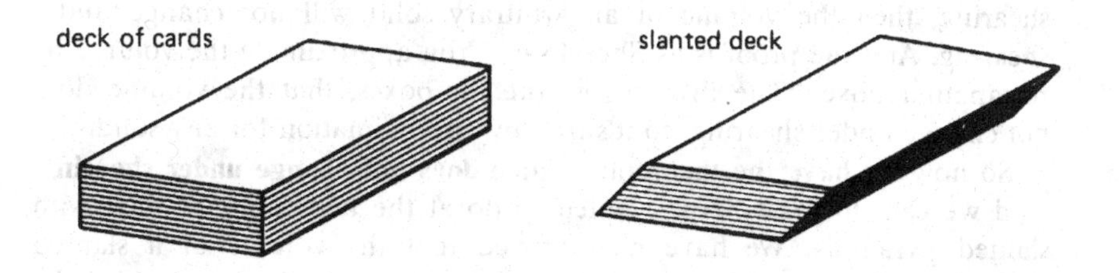

So you see the deck of cards, and over there is the sheared deck. The top is moved over like that. So if you shear a deck of cards, the volume of the deck of cards won't change. We proved for rectangles that if you shear in one direction, then the area does not change. Then in three dimensions we get the same type of theorem:

In three dimensions the volume does not change under shearing.

We have done this for rectangular boxes, shearing in one dimension.

But now we can do it to pyramids, with a square base, or a rectangular base. And it's a straight pyramid. And I shear it like that, so the height still remains the same. The perpendicular height.

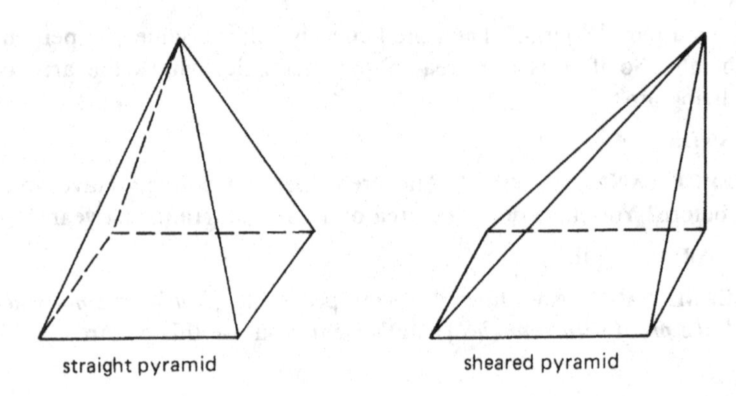

Then the volume will not change under shearing. And this holds for any 3-dimensional body. How do I prove it? Suppose I take any 3-dimensional body. I claim that its volume does not change under shearing. How do I prove it?

SHERYL. You approximate it with regular figures. Like a cube.

SERGE LANG. That's right! Perfect. You guys are very good, you know. I make a grid inside my 3-dimensional figure, a grid of rectangular boxes. Then I make a shearing. Then each rectangle goes into–what? A 3-dimensional parallelogram. It's called a parallelepiped. You know the word? If I shear a rectangular box, then this thing is called a parallelepiped. So if the volume of a rectangular box does not change under shearing, then the volume of an arbitrary solid will not change under shearing. And the proof is as Sheryl said. You approximate the volume by rectangular boxes. The theorem is true for boxes, that the volume does not change under shearing, so it's true by approximation for any solid.

So now we have the theorem: **volume does not change under shearing.** And we can do what Serge wanted to do at the beginning, to deal with slanted pyramids. We have now proved that the volume of a slanted pyramid is the same as the volume of the straight pyramid, one third the base times the height. If I have a pyramid with square base, or rectangular base, and the base is B, then

$$\textbf{Volume of slanted pyramid} = \frac{1}{3}abh = \frac{1}{3}Bh.$$

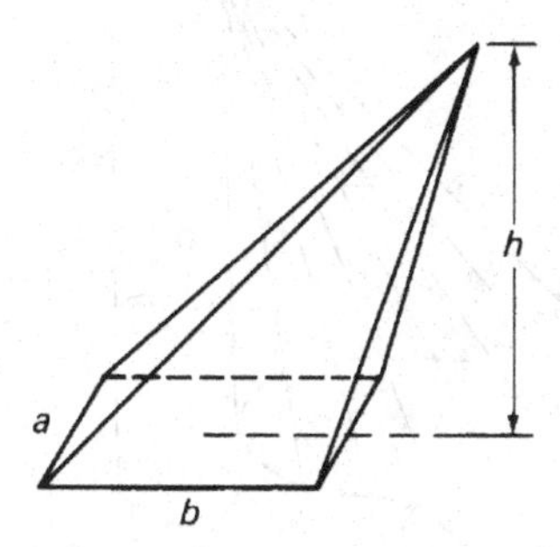

SHERYL. What about a triangular base pyramid?

SERGE LANG. Very good! But why stop at triangles? In fact why not a curved base pyramid, a pyramid with any base? The pyramid is not the only thing I want. You see the cone here, with a circle base?

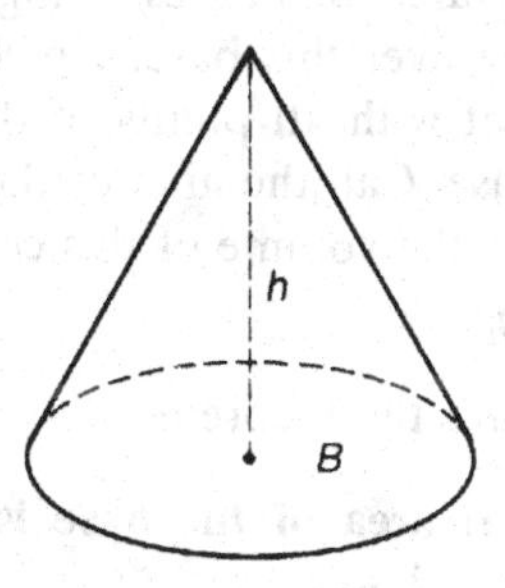

It has a base B and a certain height h. What's the volume of the cone? Who can guess? [*Several hands go up.*]

CHARLIE. It will be πr^2, times the height, times [*pause*] a third?

SERGE LANG. Yes! You got it just right. Perfect.

$$\textbf{The volume of the cone is } \frac{1}{3}\pi r^2 h.$$

[*To another student.*] What would you have said?

ANOTHER STUDENT. Same thing.

SERGE LANG. And Mike back there?

MIKE. Same thing.

SERGE LANG. That's right. That's the theorem. And suppose you had an arbitrary base, with a kidney, and I form the cone, like this.

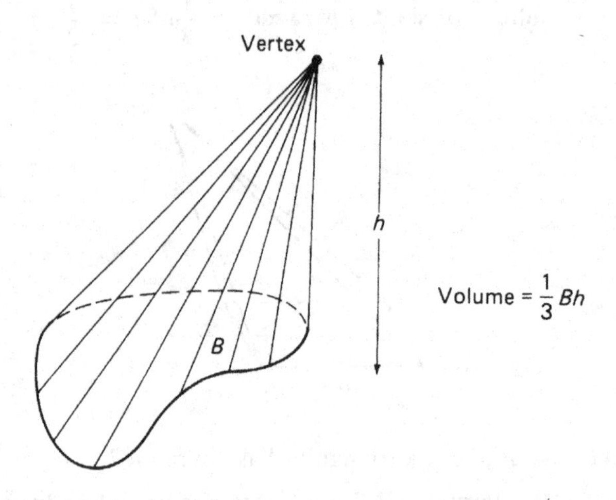

I can make the base as cockeyed as you want. I take an arbitrary base. Sometimes it might be square, sometimes it might be curved, or whatever. Then I can form the cone over this base. I pick one point somewhere in space, and I join the point with all points of the base. This gives me the cone over the arbitrary base. Call the area of the base B. Let h be the perpendicular height. What is the volume of this cone? What's the volume?

SHERYL. One third Bh.

SERGE LANG. Yes, that's the theorem:

For any cone such that area of the base is B and the perpendicular height is h, the volume is $\frac{1}{3}Bh$.

How do we prove it?

SERGE. Start with the simplest case?

SERGE LANG. The simplest case was that of a pyramid with square base, or rectangular base, and a given height. Now suppose you have an arbitrary curved base. How can you deal with an arbitrary curved base? What do I do?

SERGE. Oh, I see. You try to fit other figures by means of a grid.

SHERYL. You approximate the base with a grid.

SERGE LANG. I put a rectangular grid on the base, that's right. Then I start drawing all these lines

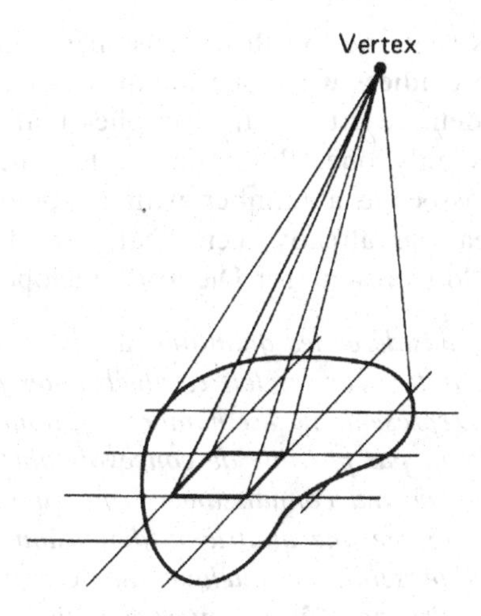

RACHEL. You approximate the volume by approximating the base with rectangles. Then you approximate the cone by pyramids. Then you can put it together to get the volume.

SERGE LANG. That's right. And I guess Sheryl would have said the same thing. [*Sheryl nods.*] And I guess Michael back there? [*Laughter.*] You see, for each rectangle of the grid, you can form the pyramid with this rectangle as the base, with the same vertex as the cone. You do this by drawing all the lines from the vertex to the points of the grid. Then you form all slanted pyramids with the vertex. Then you know that for each of these slanted pyramids, the volume is one third the base times the perpendicular height. So if you take their sum, and use the theorem for each of the slanted pyramids, you get the theorem for the whole cone, with an arbitrary base. So that's how we get $\frac{1}{3}Bh$.

Now if the base is a circle of radius r, then the area of the base is πr^2. That's how we get $\frac{1}{3}\pi r^2 h$ for the volume of the ordinary cone. It's just a very special case of the general theorem with an arbitrary base.

Comments

The ready acceptance of n dimensions by students at an early age is a common experience for me. Given the history of "dimensions", and all the fuss made around Einstein's "fourth dimension" as time, this might be surprising to some people. But on the basis of experience, I find that young people will find problems about higher dimensions only if grown ups suggest the existence of problems.

The question: "Is time the fourth dimension?" is a very bad question, because it already prejudices what one might mean by "dimension", and also the use of the definite article "the" implies that if there is a "fourth dimension" there is only one. But today it is generally accepted that whenever you can associate a number with a notion, then you have a dimension. This idea was already seen clearly by d'Alembert, when he wrote the article on "dimension" for Diderot's enclopoedia:

> *Cette manière de considérer les quantités de plus de trois dimensions est aussi exacte que l'autre; car les lettres algébriques peuvent toujours être regardées comme représentant des nombres, rationnels ou non. J'ai dit plus haut qu'il n'était pas possible de concevoir plus de trois dimensions. Un homme d'esprit de ma connaissance croit qu'on pourrait cependant regarder la durée comme une quatrième dimension et que le produit du temps par la solidité serait en quelque manière un produit de quatre dimensions. Cette idée peut être contestée mais elle a, ce me semble, quelque mérite, quand ce ne serait que celui de la nouveauté.*

Which, translated in English, reads:

> *This way of considering quantities in more than three dimensions is just as exact as the other one; because algebraic letters can always be regarded as representing numbers, rational or not. I said above that it was not possible to conceive more than three dimensions. A clever gentleman with whom I am acquainted believes that nevertheless, one could regard duration as a fourth dimension, and that the product of time by solidity would be in some way a product of four dimensions. This idea can perhaps be challenged, but it has, it seems to me, some merit, were it only that of being new.*

Of course, one of his friends is presumably himself, but he is prudent, considering what must have been at the time a rather far out idea. Today, the idea is current. In physics, you have dimensions besides the three spatial dimensions and time, for instance speed, acceleration, curvature, etc. In economics, you can associate a number, say the total gross profits of some company in any given year, so you have a dimension for, say, the steel industry, oil industry, fisheries, agriculture, automobile industry, and whatnot. In that way you can have as many dimensions as you can think of.

When I brought up higher dimensions with the students during my talk, there was absolutely no resistance from anyone. The students saw right away that the important thing is how we operate with numbers in the context of several dimensions. Then they have the correct idea concerning this abstraction, and give the right answers. Sometimes in similar circumstances, I do get the question whether a fourth dimension exists. I answer it as above, pointing out that it all depends on how you wish to use the word. If by "dimension" you want to mean only the spatial dimensions of ordinary space around you, then there are only three dimensions. If you accept to give the word a wider meaning, then you have arbitrarily many dimensions, with which you can work just as easily. The automatic response that in n dimensions under a dilation by a factor of r volume changes by a factor r^n is then both immediate and correct. Of course, when I ask how volume changes under dilations in four dimensions, Serge answers: "It's *rst* whatever." He shows that he has well understood what's going on. The only thing that would remain to be done is to point out that his "whatever" corresponds to a choice of letters, and that if we continue to use letters for dimensions, we shall run out of choices after the 26 letters of the alphabet have been exhausted. Therefore, one may denote the numbers associated with three dimensions by something like x_1, x_2, x_3; and then there is no difficulty to denote the numbers associated with n dimensions by $x_1, x_2, \ldots, x_n$. Then if we dilate by factors of $r_1, r_2, \ldots, r_n$ one sees at once that volume of boxes changes by a factor of $r_1 r_2 \cdots r_n$ (the product). Whatever difficulty exists here (and it is very slight) lies only in an appropriate choice of letters and indices; that is, in the choice of notation to transcribe in symbols what the mind has already grasped.

Concerning the definition of an ellipse, I asked the class if they knew about coordinates, and they did not. To a slightly older class, I would expand a little as follows. Traditionally, ellipses, parabolas, and other standard curves are defined by what I regard as complicated properties, for instance: the ellipse is the locus of points such that the sum of the distances from two given points is constant. I object very much to this traditional approach, which makes it difficult first to see the standard equation

$$\frac{x^2}{a^2} + \frac{y^2}{b^2} = 1,$$

exhibiting the fact that the ellipse is a dilated circle; and second, makes it difficult to see that the area is πab, which is immediate from the definition by means of the notion of dilation, and the technique of approximation by rectangles.

The volume of the ball

This lecture was given in two hours, one before lunch and one after lunch, to a class of students in Paris at about the 9th grade level, so 14 years old. They had not yet covered Pythagoras' theorem in class, for instance.

SERGE LANG. What do you do in this class? Geometry? Algebra?

THE CLASS. Yes, algebra.

SERGE LANG. And geometry?

THE CLASS. Yes, a little.

SERGE LANG. So you know some geometry?

THE CLASS. We are just beginning.

SERGE LANG. [*Pointing to a student.*] You know what π is?

STUDENT. Yes.

SERGE LANG. So, what is it?

STUDENT. It's a number . . . geometric.

SERGE LANG. Oh yes? Which one?

STUDENT. 3.14 and so on.

SERGE LANG. 3.14 and so on. And what does π represent?

STUDENT. ???

SERGE LANG. OK, who knows what π represents? You, what's your name?

STUDENT. Christopher.

SERGE LANG. All right, Christopher. So?

NATHALIE. The circumference of a circle.

SERGE LANG. So what is the circumference of a circle?

NATHALIE. What do you mean, what is it?

SERGE LANG. I mean, as a function of π.

NATHALIE. ???

SERGE LANG. Well, you have a circle of radius r . . .

A STUDENT. Ah! It's $2\pi r$.

SERGE LANG. Very good. $2\pi r$, where r is the radius. And the surface, the area, what is it?

THE STUDENT. It's πr squared.

SERGE LANG. That's right. So we have two formulas, and as Christopher said, π is equal to 3.14 . . . . Good, I am not going to do this, but I want to do the same thing in one higher dimension. If we go to a higher dimension . . . you see, before we had the circumference and the area, but if we have one more dimension, what do we get?

THE STUDENT. A ball.

SERGE LANG. Right, a ball. I draw it.

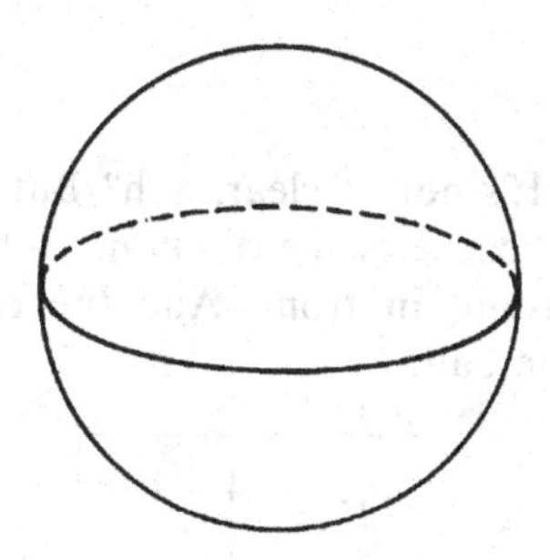

And what do we want to find about the ball?

THE STUDENT. The volume.

SERGE LANG. That's right, the volume, What's your name?

STUDENT. Ann.

SERGE LANG. You know the formula for the volume of the ball of radius r?

ANN. ???

SERGE LANG. Who knows the formula?

THE CLASS. ???

SERGE LANG. All right. The length of the circle is $2\pi r$ and the area is πr^2. But for the ball, what is it going to be?

A STUDENT. π squared times r to the power 4.

SERGE LANG. No. So let's go back to boxes. Suppose I have a square with sides equal to r. Then its area is equal to what?

THE CLASS. r squared.

SERGE LANG. Now, suppose I have a cube. Its volume is what?

THE CLASS. r cube.

SERGE LANG. Have you done some dilations?

THE CLASS. A little.

SERGE LANG. Good. Let's take a cube with side a. Then the volume is a^3. If I make a dilation by a factor of r in all directions, then the volume becomes what? Ann.

ANN. ra quantity cubed.

SERGE LANG. That's right, $(ra)^3$. In other words, r^3a^3. OK? Everybody understood? So we can say that if I make dilation by a factor of r, then the volume changes, how? By a factor of r^3. OK? Good. Now let's go on to the sphere. The volume is going to be what?

SOPHIE. πr cube.

SERGE LANG. πr^3–it's not at all stupid to say that. And you, what do you say?

ANN. π cube r cube.

SERGE LANG. Ah! It's not so clear, heh? But there must be the factor of r^3 no matter what, because of the dilation. So the answer is that there is a πr^3, but with a constant in front. And the constant is equal to 4/3. That's the volume of the ball:

$$\boxed{V = \frac{4}{3}\pi r^3.}$$

Do you know the volume of other geometric figures? Or do you just know surfaces? Is this the first volume that you have ever seen?

THE CLASS. No, there is the cube.

SERGE LANG. All right. So what I want to do today is to derive the volume of the ball: $\frac{4}{3}\pi r^3$. I want to prove it. How are we going to prove it? Nobody knows? First we are going to find the volume of the ball of radius 1. It's simpler. What is the volume of the ball of radius 1? Ann.

ANN. ???

SERGE LANG. If I let $r = 1$ in the formula, what do I get?

ANN. $\frac{4}{3}\pi$.

SERGE LANG. That's right, $\frac{4}{3}\pi$. We want to prove that the volume of the ball of radius 1 is $\frac{4}{3}\pi$. And how do we do it?

THE CLASS. ???

SERGE LANG. Well, we have to start from things we know, for instance cylinders, or boxes.

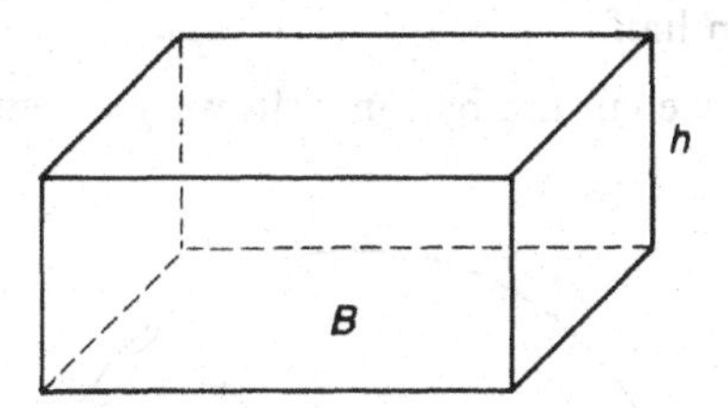

The volume of a rectangular box of base B and height h is equal to what?

SOPHIE. The base times the height.

SERGE LANG. That's right, B times h. Here I took a base which was a rectangle. But suppose I took a curved base? Any base, like this for instance.

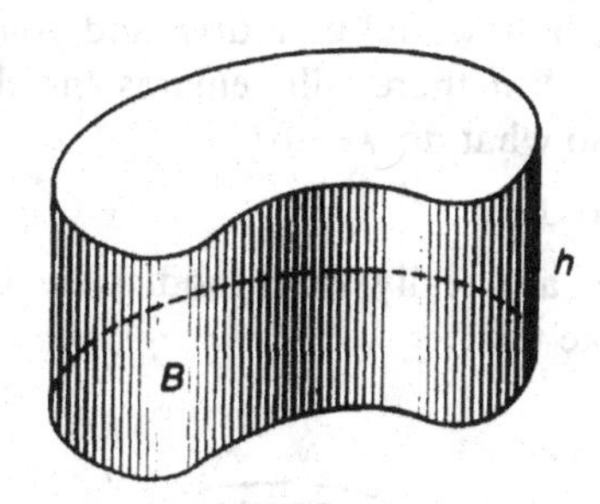

THE CLASS. It's the same thing.

SERGE LANG. Yes, it's the same thing, it's always the base times the height. And if I have a real cylinder, whose base is a disc of radius r, and height h? You. [*Pointing to a student.*]

VERONIQUE. ???

SERGE LANG. The area of the base is what?

VERONIQUE. ???

SERGE LANG. It's a disc, the base is a disc. The area of a disc is what?

RIVKA. πr^2.

SERGE LANG. And the volume of the cylinder, it's equal to what?

THE CLASS. $\pi r^2 h$.

SERGE LANG. That's right:

Volume of cylinder of radius r and height h = $\pi r^2 h$.

OK? If you know the area of the disc, πr^2, then the volume of the cylinder is $\pi r^2 h$. Everybody agrees? Good. Now we come to the ball. What do we do? We have to cut up the ball into pieces that we know, and for the moment, what we know are cylinders.

SOPHIE. Cut it in half.

SERGE LANG. If we cut the ball in half, we get something like this.

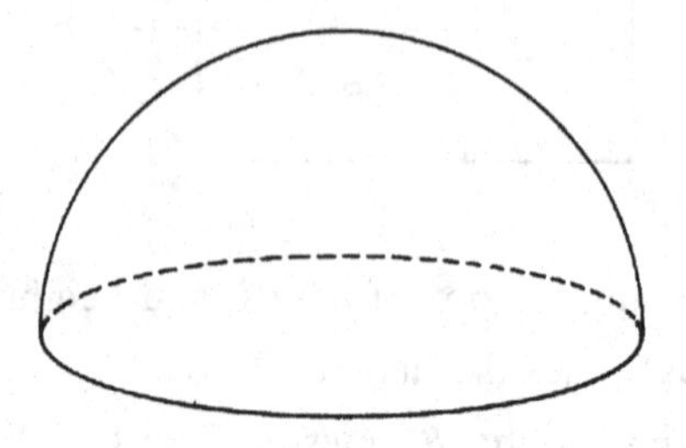

SOPHIE. We could cut it up in slices, the base would be a disc.

SERGE LANG. The base would be a disc, and you know the area of the base, πr^2. That's right. But there still remains the thing on top, which is not quite a cylinder. So what do we do?

ANN. Cut it up into discs.

SERGE LANG. You are really very good. [*Laughter.*] That's right, we cut it up into slices, like this:

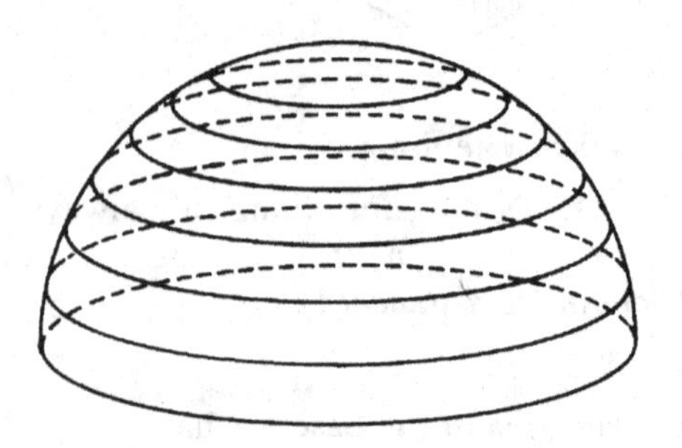

Good. Do you know Pythagoras' theorem?

THE CLASS. No.

SERGE LANG. I have a right triangle, like this:

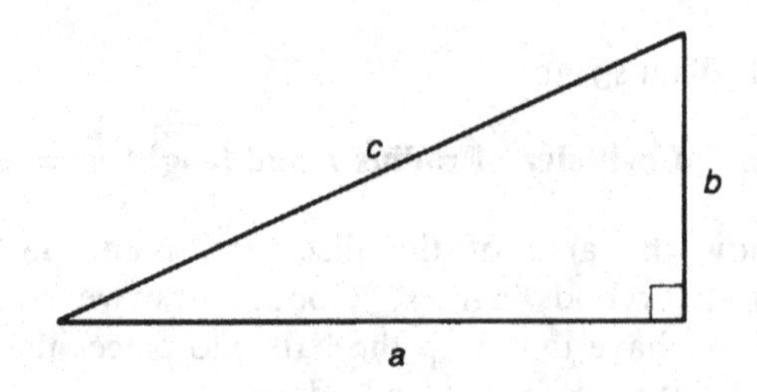

One side is a, another side is b, and the third side, the hypotenuse, is c. You don't know the relation between $a,b,$ and c? You have never seen it? We have

$$a^2 + b^2 = c^2.$$

You've never seen this?

THE TEACHER. Next year.

A STUDENT (VANYA). Why the squares?

SERGE LANG. It's not at all obvious why you need the squares. One has to prove it, but it's not what I wanted to do today. So I am asking you to accept it. Can you accept it?

THE CLASS. Yes, OK.

SERGE LANG. Good. Now I cut up the ball and I make cylindrical slices, like this.

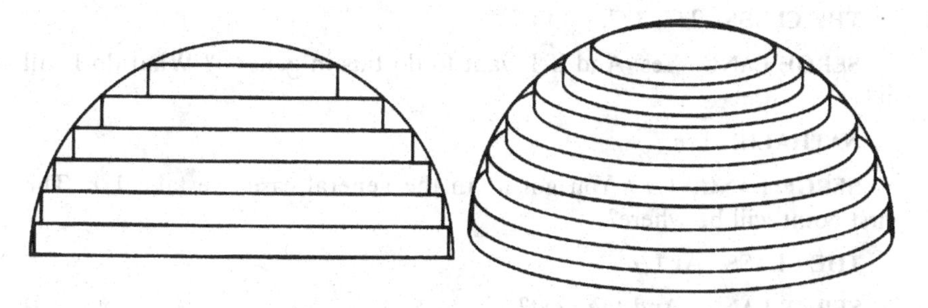

Actually, I took a half ball, and I cut it up. Does everybody see the slices? If we look at them sideways, they look like rectangles. But if I turn them around the vertical axis, then I get cylinders. And I know the volumes of these cylinders once I know their radius and their height. It's what?

STUDENTS. $\pi r^2 h$.

SERGE LANG. OK. Now I need to know the radius and the heights. Well, the heights will depend on the number of cylinders. I can have 5, or 6, or more. Let's draw the picture with six.

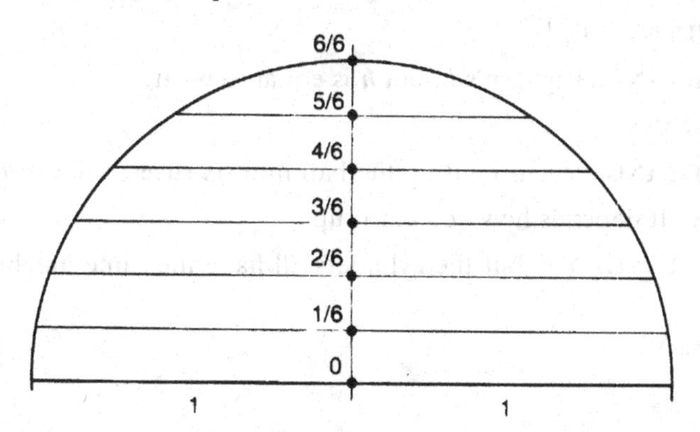

If I call the origin 0, then the first point is at what height?

THE CLASS. 1/6.

SERGE LANG. And the next?

THE CLASS. [*All together.*] 2/6.

SERGE LANG. And the next?

THE CLASS. 3/6.

SERGE LANG. That's right. And the next ones have height 4/6, 5/6 and 6/6.

THE CLASS. Yes.

SERGE LANG. And if I had seven slices? What would be the height of the first point?

THE CLASS. 1/7.

SERGE LANG. And the others?

THE CLASS. 2/7, 3/7, up to 7/7.

SERGE LANG. Yes. And if I want to do this in general? What do I call it?

NATHALIE. $1/n$.

SERGE LANG. $1/n$? You got it! In the general case, we take $1/n$. The first point will be where?

THE CLASS. At $1/n$.

SERGE LANG. And the next?

THE CLASS. At $2/n$.

SERGE LANG. And the next?

THE CLASS. At $3/n$.

SERGE LANG. And the last one, all the way at the end?

THE CLASS. n/n.

SERGE LANG. That's right, that's what happens in general. And what is the volume of each cylinder going to be? The height is equal to what?

A STUDENT. It's h.

SERGE LANG. Right, it's h, but h is equal to what?

THE CLASS. ...

SERGE LANG. When I cut up the ball into six slices, what is h?

VANYA. It depends how you cut it up.

SERGE LANG. Yes, but the cylinders all have the same height. Do you see that?

VANYA. Yes.

SERGE LANG. What is the height, if I have six slices?

VANYA. 1/6.

SERGE LANG. Yes, when I have six slices, then the first height is 1/6; and the second height is also 1/6. And the others also. OK?

CLASS. Yes.

SERGE LANG. All right, then in the case of six slices, all the heights are equal to 1/6. In the case of n slices, these heights are equal to what?

THE CLASS. [*All together.*] 1 over n.

SERGE LANG. Very good, $1/n$. Now the radius, what is the radius equal to? The radius of the first slice?

THE CLASS. r.

SERGE LANG. Yes, it's r, but r is equal to what? OK, I draw it with six. There is a first r, then a second r, and so forth until the last one.

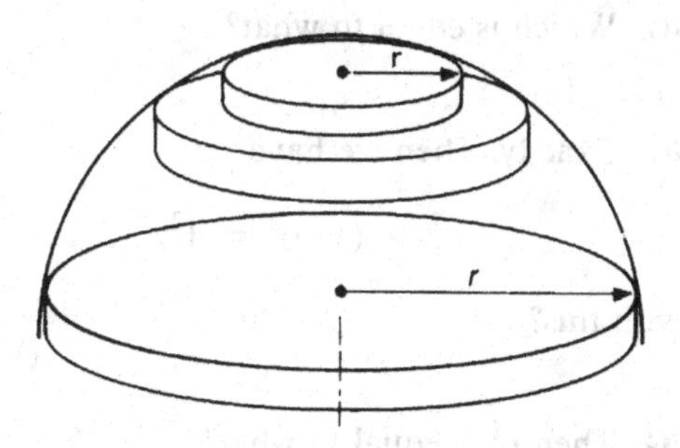

The radii change, don't they? Then the first radius is equal to what? It's the radius of the big circle, the radius of the ball, which I take to be equal to 1. Remember, we are trying to figure out the volume of the ball of radius 1.

And the radius of the first cylinder, it's OA_1. It's equal to what?

THE CLASS. ???

SERGE LANG. We don't quite know, it's smaller than 1. Agreed?

THE CLASS. Yes.

SERGE LANG. Well, here is where we need Pythagoras' theorem. Let's draw the first cylinder.

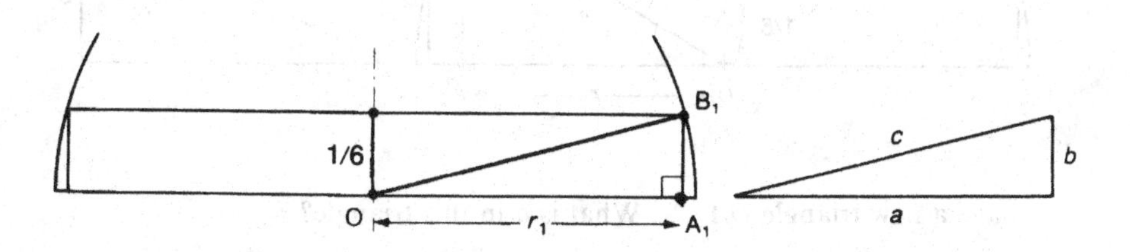

We have a right triangle OA_1B_1 in the picture. And in a right triangle, we have agreed before that

$$a^2 + b^2 = c^2.$$

In the right triangle OA_1B_1, what is a?

THE CLASS. It's the radius of the cylinder.

SERGE LANG. That's right, and the radius of the cylinder is r_1. And what is b?

A STUDENT (PHILIP). It's the height.

SERGE LANG. And the height is equal to what?

PHILIP. 1/6.

SERGE LANG. Right. And c is equal to what?

PHILIP. It's the radius of the ball.

SERGE LANG. Which is equal to what?

PHILIP. Euh . . . 1.

SERGE LANG. Exactly. Then we have

$$r_1^2 + (1/6)^2 = 1^2.$$

Do you agree with this?

PHILIP. Yes.

SERGE LANG. Then r_1^2 is equal to what?

PHILIP. $1 - (\frac{1}{6})^2$,

SERGE LANG. That's right, very good. Now let's look at the second cylinder. Let me draw it.

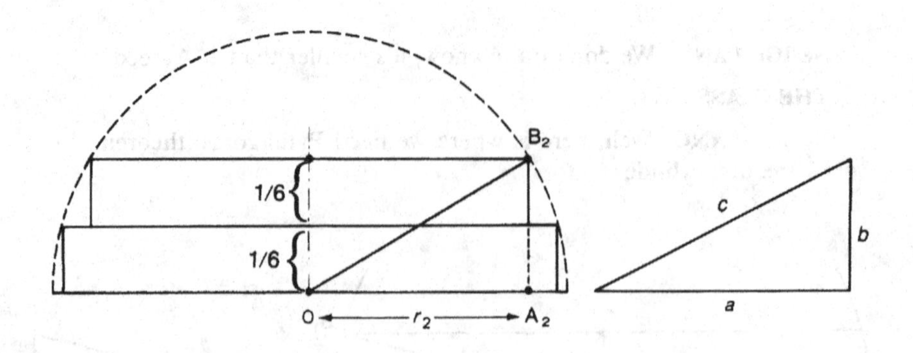

I have a new triangle OA_2B_2. What is a in this triangle?

PHILIP. It's the radius of the cylinder.

SERGE LANG. Yes, the radius of the second cylinder, r_2. And b?

PHILIP. It's A_2B_2.

SERGE LANG. Which is equal to . . . ?

PHILIP. 2/6.

SERGE LANG. Very good. And c is still equal to 1. Then what is r_2^2?

ANOTHER STUDENT. Then r_2^2 will be equal to $1^2 - (2/6)^2$?

SERGE LANG. There you are. And the next one will be r_3, and we have $r_3^2 = ?$

THE STUDENT. $r_3^2 = 1 - (\frac{3}{6})^2$.

SERGE LANG. That's right. OK? We can go on? We go on like this with r_4, r_5, r_6. I write it down [*and the students repeat out loud*]:

$$r_1^2 = 1 - (\frac{1}{6})^2$$

$$r_2^2 = 1 - (\frac{2}{6})^2$$

$$r_3^2 = 1 - (\frac{3}{6})^2$$

$$r_4^2 = 1 - (\frac{4}{6})^2$$

$$r_5^2 = 1 - (\frac{5}{6})^2$$

$$\text{and } r_6^2 = 1 - (\frac{6}{6})^2.$$

Now we have some idea about the radii of the cylinders. And the height of these cylinders, what is it?

A STUDENT (KARIN). It depends. 1/6, 2/6 . . .

SERGE LANG. No, that was for the triangles, when we computed the radii. But the cylinders all have the same height. And it's equal to what?

THE STUDENT. 1/6.

SERGE LANG. That's right, very good, 1/6. And the radii of the cylinders depend on the cylinders. The radii are going to be different. The first radius is r_1. The second radius is r_2. The third is r_3. They become smaller and smaller. OK? Now, the volume of the cylinder is what?

A STUDENT (RAJNA). The base times the height.

SERGE LANG. Yes, it's $\pi r^2 h$. The radius of the first cylinder is r_1. And the volume of the first cylinder is . . .

SOME STUDENTS. [*All together*]. πr_1^2 times 1/6.

SERGE LANG. And the volume of the second cylinder is what?

THE STUDENTS. πr_2^2 times 1/6.

OTHERS. πr_2^2 times 2/6.

SERGE LANG. Who said 1/6?

[*Some hands go up.*]

SERGE LANG. Who said 2/6?

[*Other hands go up.*]

SERGE LANG. No, no! The height of the second cylinder is still 1/6. OK? Then the volume of the second cylinder is what? It is

$$V = \pi r_2^2 \frac{1}{6}.$$

And the volume of the third cylinder?

RAJNA. It's $\pi r_3^2 \frac{1}{6}$.

SERGE LANG. That's right. And the next, Ann?

ANN. πr_4^2 times $\frac{1}{6}$.

SERGE LANG. Yes. But the squares of the radii r_1, r_2, we have already found them a minute ago. What are these squares equal to?

NORA. $1 - (\frac{1}{6})^2, 1 - (\frac{2}{6})^2, \ldots$

SERGE LANG. So what do I get when I take the sum of the volumes of all these cylinders? You, what's your name?

A STUDENT. Siem Lang.

SERGE LANG. Gee, like me! [*Laughter.*] OK, so I take the sum [*and Serge Lang writes as he questions the students*]. What's the first one, Siem Lang?

SIEM LANG. $\pi r_1^2 \cdot \frac{1}{6}$.

SERGE LANG. And r_1^2 is equal to what?

SIEM LANG. $1 - (\frac{1}{6})^2$.

SERGE LANG. OK. Therefore the volume of the first cylinder is

$$\pi(1 - (\frac{1}{6})^2) \times \frac{1}{6}.$$

What about the second?

SIEM LANG. $\pi(1 - (\frac{2}{6})^2) \times \frac{1}{6}$.

SERGE LANG. And the third?

SIEM LANG. $\pi(1 - (\frac{3}{6})^2) \times \frac{1}{6}$.

SERGE LANG. Good. Then we go on, and we find that the sum of the volumes is equal to:

sum of the volumes of the cylinders =

$$\pi(1 - (\frac{1}{6})^2) \cdot \frac{1}{6} + \pi(1 - (\frac{2}{6})^2) \cdot \frac{1}{6} + \cdots + \pi(1 - (\frac{6}{6})^2) \cdot \frac{1}{6}.$$

Now, suppose I took seven cylinders instead of six. Then what would be the sum? You, what's your name?

A STUDENT. Cyrille.

SERGE LANG. OK, Cyrille, what's the sum?

CYRILLE. You must replace the six by seven. It's

$$\pi(1 - (\frac{1}{7})^2) \cdot \frac{1}{7}.$$

SERGE LANG. Very good. And it will continue up to what?

CYRILLE. $\pi(1 - (\frac{7}{7})^2) \cdot \frac{1}{7}$.

SERGE LANG. Now, we could take 6, 7, 8, 9, 10, . . . What if I just take n, what do I get for the volume of the first cylinder?

CYRILLE. $\pi(1 - (\frac{1}{n})^2) \cdot \frac{1}{n}$.

SERGE LANG. Fantastic! And the next? Someone else.

A STUDENT. $\pi(1 - (\frac{2}{n})^2) \cdot \frac{1}{n}$.

SERGE LANG. Yes, it goes on like this. What about the last? [*Pointing.*] What's your name?

THE STUDENT. Kalasa. $\pi(1 - (\frac{n}{n})^2) \cdot \frac{1}{n}$.

SERGE LANG. That's right. So the sum of the volumes of all the cylinders is:

sum of volumes =

$$\pi(1 - (\frac{1}{n})^2) \cdot \frac{1}{n} + \pi(1 - (\frac{2}{n})^2) \cdot \frac{1}{n} + \cdots + \pi(1 - (\frac{n}{n})^2) \cdot \frac{1}{n}.$$

Now, when n gets bigger and bigger, what happens to this sum? It gets close to what?

THE CLASS. ???

SERGE LANG. I take more and more cylinders, like this.

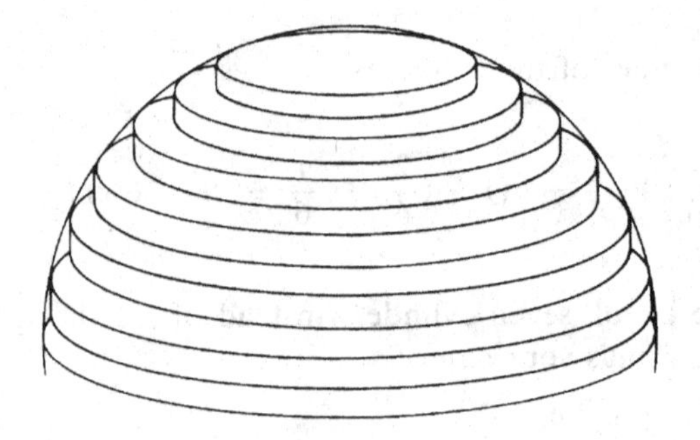

What does the sum of the volumes of the cylinders approach?

PHILIP. The volume of the ball.

SERGE LANG. [*Writing:*]

When n becomes larger and larger, the sum of the volumes of the cylinders approaches the volume of the half ball of radius 1.

So now, the problem becomes: what happens to the expression on the right when n becomes bigger and bigger?

ANN. You can compute it.

SERGE LANG. You want to compute it directly? OK, but how?

ANN. Compute the volumes one after the other.

SERGE LANG. Yes, it works if you have any specific number, like six, or ten. But what about n? When n becomes bigger and bigger? You don't know any specific value for n. So what we now have to do is to figure out what happens to the sum when n gets bigger and bigger. But we'll do this after lunch. See you this afternoon.

The second hour

[*The class returns after lunch.*]

SERGE LANG. So we're back, and the problem was to find the volume of the ball of radius 1. We had derived the expression for the volume of the half ball, which is approached by

$$\pi(1-(\frac{1}{n})^2)\cdot\frac{1}{n}+\pi(1-(\frac{2}{n})^2)\cdot\frac{1}{n}+\cdots+\pi(1-(\frac{n}{n})^2)\cdot\frac{1}{n}$$

when n becomes bigger and bigger. We can already factor out π, and also factor out $\frac{1}{n}$, so we can write the expression as

$$\frac{\pi}{n}[1-(\frac{1}{n})^2+1-(\frac{2}{n})^2+1-(\frac{3}{n})^2+\cdots+1-(\frac{n}{n})^2].$$

Now let's look at the big parenthesis. We see some one's, 1, then 1, then 1, . . . How many one's are there?

THE CLASS. n.

SERGE LANG. Yes. So I get

$$\frac{\pi}{n}(n - \text{something to be subtracted}).$$

What do we subtract?

RAJNA. $(1/n)^2$.

SERGE LANG. That's right. And after that?

RAJNA. $(2/n)^2$.

SERGE LANG. Very good. And the next, it's what? Christopher?

CHRISTOPHER. $(3/n)^2$.

SERGE LANG. Right, up to $(n/n)^2$. So we subtract

$$(\frac{1}{n})^2+(\frac{2}{n})^2+(\frac{3}{n})^2+\cdots+(\frac{n}{n})^2.$$

And if we write it all down together, then what's the full expression, Christopher?

CHRISTOPHER. $\pi\frac{n}{n}-\frac{\pi}{n}[(\frac{1}{n})^2+(\frac{2}{n})^2+\cdots+(\frac{n}{n})^2].$

SERGE LANG. But $\pi n/n$ is equal to what?

CHRISTOPHER. π.

SERGE LANG. Yes. Now subtract . . .

NATHALIE. $\frac{\pi}{n}$ . . . euh . . .

SERGE LANG. Yes, go on.

NATHALIE. Times $(1/n)^2 + (2/n)^2 + (3/n)^2 + \cdots$

SERGE LANG. Very good, up to $(n/n)^2$. So the formula now becomes

$$\pi - \pi\left[\left(\frac{1}{n}\right)^2 \cdot \frac{1}{n} + \left(\frac{2}{n}\right)^2 \cdot \frac{1}{n} + \left(\frac{3}{n}\right)^2 \cdot \frac{1}{n} + \cdots + \left(\frac{n}{n}\right)^2 \cdot \frac{1}{n}\right],$$

and this expression approaches the volume of the half ball when n gets bigger and bigger. Now, if the formula I gave at the beginning is correct, then what does the expression in parentheses approach?

THE CLASS. ???

SERGE LANG. The volume of the half ball of radius 1 is supposed to be equal to $\frac{2}{3}\pi$. Here we have the expression

$$\pi - \pi\left[\left(\frac{1}{n}\right)^2 \frac{1}{n} + \cdots + \left(\frac{n}{n}\right)^2 \frac{1}{n}\right].$$

If this expression approaches two thirds of π, then what does the sum in the parentheses approach? Do you see it? I have a π, and I subtract a certain fraction of π. I want what remains to be two thirds of π. Then the fraction is what?

A STUDENT. $1/3$.

SERGE LANG. That's it! One third. What's your name? Elizabeth? She got it, Elizabeth got it. The expression in parentheses approaches $1/3$. It's very good. And that's what we have to show. And if we succeed, then we win. OK, let's write this down. Up to now, we have shown that

the sum of the volumes of the cylinders is equal to

$$\pi - \pi\left[\left(\frac{1}{n}\right)^2 \cdot \frac{1}{n} + \left(\frac{2}{n}\right)^2 \cdot \frac{1}{n} + \cdots + \left(\frac{n}{n}\right)^2 \cdot \frac{1}{n}\right],$$

and now, we want to prove that the sum

$$\left[\left(\frac{1}{n}\right)^2 \cdot \frac{1}{n} + \left(\frac{2}{n}\right)^2 \cdot \frac{1}{n} + \cdots + \left(\frac{n}{n}\right)^2 \cdot \frac{1}{n}\right] \textbf{ approaches } \frac{1}{3}.$$

At first sight, it looks pretty tough. There are some squares, lots of n, and it is not at all clear how the sum behaves. So what do we do?

THE CLASS. ???

SERGE LANG. Suppose there were no squares, do you think you might be able to figure it out? OK, we found an expression with squares, and we don't know what to do. The problem looks difficult, and we try to consider a similar, but simpler problem. So let's try it without the squares. Suppose I write the sum without the squares. Then it's like what? Christopher?

CHRISTOPHER.

$$\frac{1}{n}\cdot\frac{1}{n}+\frac{2}{n}\cdot\frac{1}{n}+\frac{3}{n}\cdot\frac{1}{n}+\cdots+\frac{n}{n}\cdot\frac{1}{n}.$$

SERGE LANG. That's right. Now we have a sum of fractions, whose denominator is . . . Elizabeth?

ELIZABETH. $n \cdot n$.

SERGE LANG. Yes, you can also say n square, but you are right. We can leave it as n times n. And the numerators?

ELIZABETH. $1, 2, 3, 4, \ldots, n$.

SERGE LANG. Yes. And how do I figure the sum out? Let's draw a picture. I'll do it like this. I take a square of sides 1, and I cut it up in pieces, I cut it up into little squares whose sides are $1/n$.

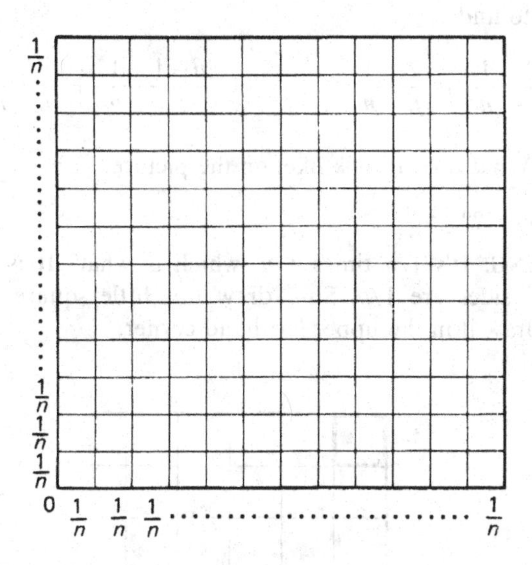

Take the point 0 in the lower left hand corner. Then if I go to the right, I get points $1/n$, $2/n$, $3/n$, $\cdots$ and the last one is what . . . Christopher?

CHRISTOPHER. n/n.

SERGE LANG. Good. And the next to the last?

NATHALIE. $(n-1)/n$.

SERGE LANG. Right, and the one before that?

ALL TOGETHER. $(n-2)/n$.

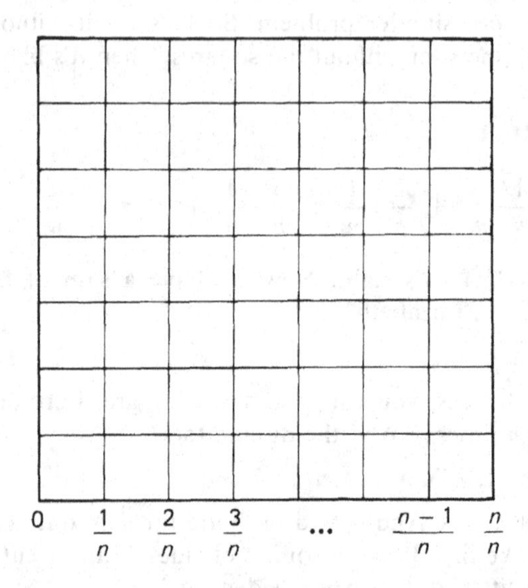

SERGE LANG. Yes, and I do the same thing vertically, up and down. Now, I want to find

$$\frac{1}{n}\cdot\frac{1}{n}+\frac{2}{n}\cdot\frac{1}{n}+\cdots+\frac{n-1}{n}\cdot\frac{1}{n}+\frac{n}{n}\cdot\frac{1}{n}.$$

Take $\frac{1}{n}\cdot\frac{1}{n}$. What does it look like, on the picture?

THE CLASS. ???

SERGE LANG. It's $1/n$ times $1/n$, which is what? It is the area of a square whose sides are $1/n$. So I draw this little square inside the big square, let's draw it in the upper left hand corner.

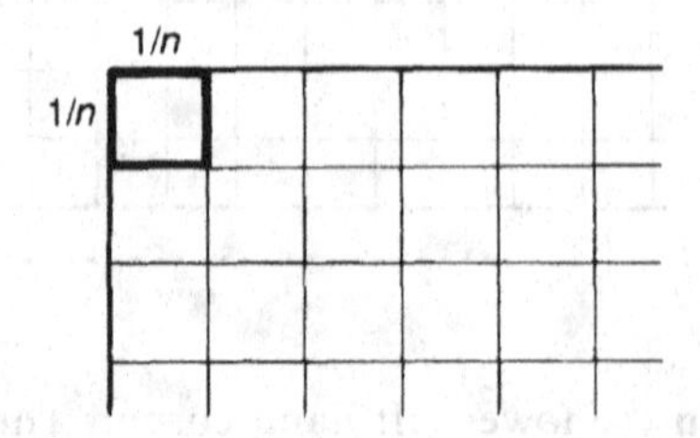

What about the next one? It's $2/n$ times $1/n$. This represents what?

A STUDENT. It's two little squares.

SERGE LANG. That's right, it's a rectangle of length $2/n$ and of width $1/n$. I draw it just below the first square.

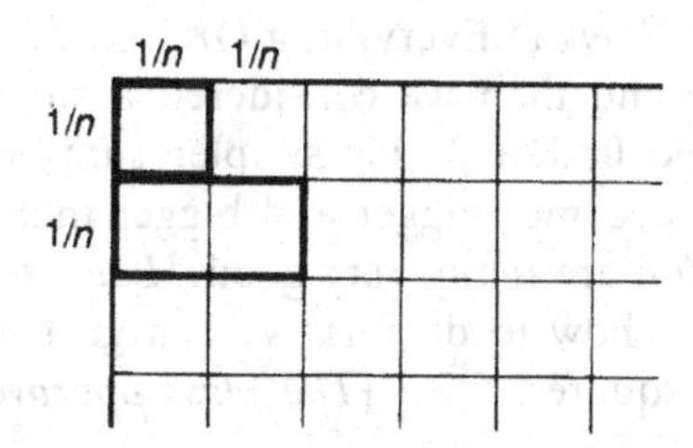

What about $3/n$ times $1/n$?

THE CLASS. It's a rectangle of length $3/n$.

SERGE LANG. And the term with $(n-1)/n$?

THE CLASS. It's a rectangle of length $(n-1)/n$.

SERGE LANG. And the last one?

THE CLASS. It's a rectangle of length n/n.

SERGE LANG. Right, of length 1, like this, all the way at the bottom. So the picture is like a staircase:

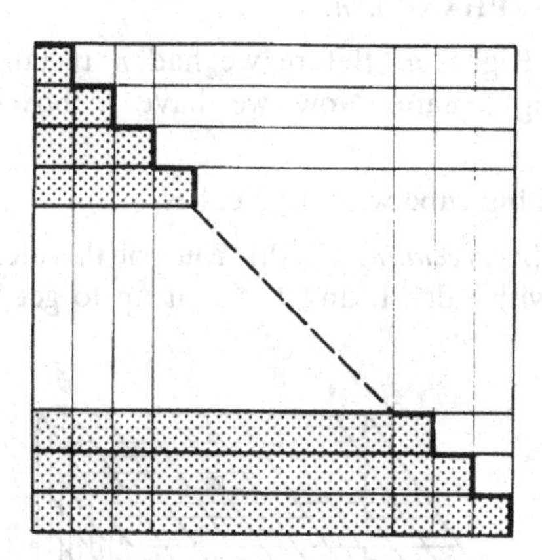

And the sum is just the area of the staircase. Now if I increase n, if n gets bigger and bigger, what does the staircase look like?

VANYA. It will have more steps.

SERGE LANG. Precisely, it will have more steps. And the area of the staircase approaches what? When n gets bigger and bigger?

THE STUDENT. It approaches half the square.

SERGE LANG. What's your name?

THE STUDENT. Rivka.

SERGE LANG. You got it, Rivka got it! Let me write it down:

The area of the staircase approaches 1/2 when *n* becomes large.

Do you all agree with Rivka? Everything OK? Good. So we started with a sum having squares, and then we considered a simpler sum without the squares. And now we find that this simpler sum is a staircase, and we showed that when n becomes bigger and bigger then the sum approaches $1/2$. That's terrific. You are really very good. [*Laughter.*]

Now that we know how to do this, we can go back to the sum at the beginning, with the squares. OK? [*The class approves.*] Here is the sum again.

$$\left(\frac{1}{n}\right)^2 \cdot \frac{1}{n} + \left(\frac{2}{n}\right)^2 \cdot \frac{1}{n} + \left(\frac{3}{n}\right)^2 \cdot \frac{1}{n} + \cdots + \left(\frac{n}{n}\right)^2 \cdot \frac{1}{n}.$$

What do we do? [*Questioning looks.*]

All right, take one term in the sum, for instance $\left(\frac{2}{n}\right)^2 \cdot \frac{1}{n}$. What is it? What does it represent? It's a volume, isn't it? If I multiply the square of something by something else, I get the volume of a box with one side equal to $1/n$ and the other two sides equal to $2/n$. Do you agree? OK. Now how many boxes are there?

A STUDENT (STEPHANIE). n.

SERGE LANG. Right, n. Before we had n rectangles, and we drew them inside a big square. Now we have n boxes. We draw them in . . . what?

STEPHANIE. A big cube with little cubes.

SERGE LANG. [*Appreciatingly*] Ah! You got the idea. That's right, so I draw a big cube with sides 1, and I slice it up to get little cubes of sides $1/n$.

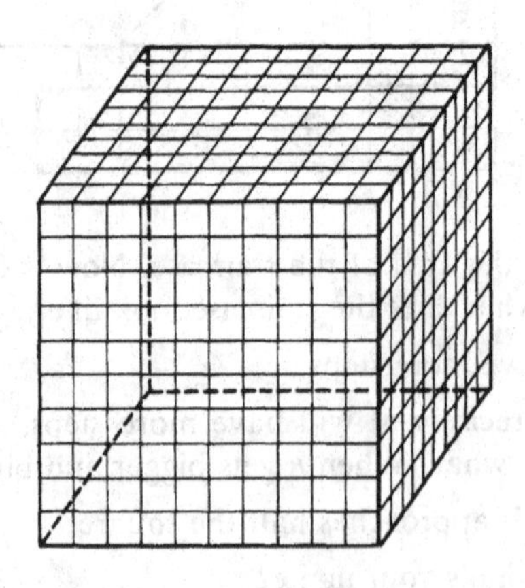

So here is the big cube. Now the first term in my sum is $(\frac{1}{n})^2 \cdot \frac{1}{n}$, and it represents what? It's the volume of a cube all of whose sides are equal to $1/n$. I draw it just like the square before.

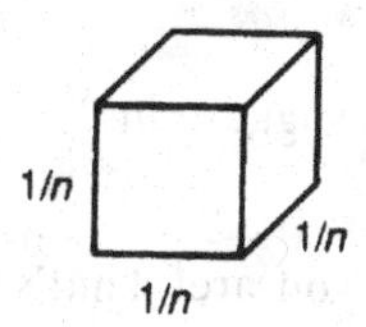

And the next one, $(\frac{2}{n})^2 \cdot \frac{1}{n}$, it's what? It's not a cube.

PHILIP. It's a flat box.

SERGE LANG. That's right, it's a flat box, with two sides equal to $2/n$, and with height $1/n$. So I draw this flat box just below the first one.

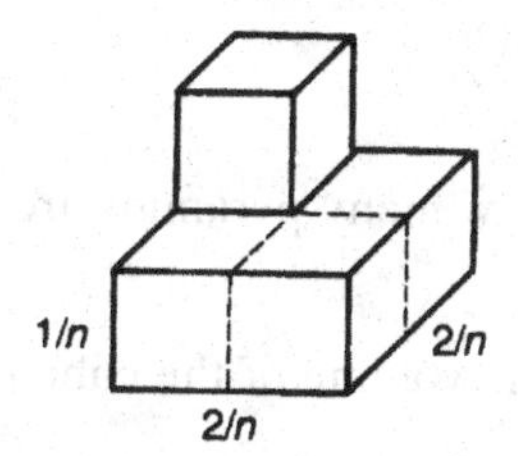

And the third, $(\frac{3}{n})^2 \cdot \frac{1}{n}$? Elisabeth?

ELISABETH. A flat box of sides $3/n$.

SERGE LANG. That's right. The height is still $1/n$, and the length of the sides is $3/n$. And we go on like this up to n/n.

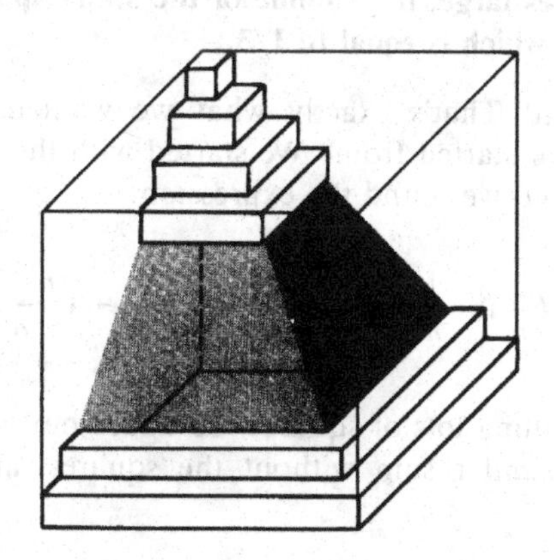

So in the sum with the squares, we also get a staircase, but these are real stairs, with a volume; they are three dimensional stairs, like those in real life. Now, when n becomes bigger and bigger, the stairs approach what?

A STUDENT. 1/3.

ANOTHER. 1/2.

SERGE LANG. Any other suggestions?

SOPHIE. A pyramid.

SERGE LANG. Ah, there you are! That's what I wanted to hear. When n becomes large, the stairs approach a pyramid. You agree with this? And how do we get this pyramid? By joining one corner, the vertex, all the way up on the left, with the opposite face on the bottom. How many faces are there which are opposite to the upper left hand corner?

RIVKA. Three?

SERGE LANG. Yes, there are three opposite faces. This means that I could have three pyramids like the one we just drew for the staircase. And these three pyramids fill out the cube. Therefore, the volume of one of the pyramids is equal to what?

VANYA. 2/3.

SERGE LANG. No. How many pyramids are then in the cube?

VANYA. Three.

SERGE LANG. And the volume of the cube is . . . ?

VANYA. 1.

SERGE LANG. Therefore the volume of the pyramid is . . . ?

VANYA. 1/3.

SERGE LANG. That's right, 1/3. Let's go back to the stairs with steps of height $1/n$. When n becomes large, then the stairs approach the pyramid, OK? So I write it down:

When n becomes large, the volume of the stairs approaches the volume of the pyramid, which is equal to 1/3.

But we have won! That's exactly what we wanted to prove. Do you remember what we started from? We started with the sum of the volumes of the cylinders, and we found the expression:

$$\pi - \pi\left[\left(\frac{1}{n}\right)^2 \cdot \frac{1}{n} + \left(\frac{2}{n}\right)^2 \cdot \frac{1}{n} + \left(\frac{3}{n}\right)^2 \cdot \frac{1}{n} + \cdots + \left(\frac{n-1}{n}\right)^2 \cdot \frac{1}{n} + \left(\frac{n}{n}\right)^2 \cdot \frac{1}{n}\right],$$

with a sum containing lots of squares. To figure out what this sum is like, we first tried a similar sum without the squares, and we found a flat staircase, with an area which approaches 1/2 when n becomes large. Now, for the sum with the squares, we find that it approaches three-dimensional stairs, and that it approaches 1/3. The sum is the volume of a staircase, which approaches 1/3 when n becomes large. Therefore the sum of the volumes of cylinders approaches

$$\pi - \frac{\pi}{3}.$$

And this is equal to what?

ELISABETH. $2\frac{\pi}{3}$.

SERGE LANG. So what we have just proved is that the volume of the half ball is equal to what? Elisabeth?

ELISABETH. $2\pi/3$.

SERGE LANG. And therefore the volume of the whole ball is . . .?

ELISABETH. $4\pi/3$.

SERGE LANG. And that's the theorem we wanted to prove. We won!

Theorem. The volume of the ball of radius 1 is equal to $4\pi/3$.

By the way, we also found out something about those sums and what they approach when n becomes large.

The first one, without the squares, is

$$\frac{1}{n}\cdot\frac{1}{n} + \frac{2}{n}\cdot\frac{1}{n} + \frac{3}{n}\cdot\frac{1}{n} + \cdots + \frac{n-1}{n}\cdot\frac{1}{n} + \frac{n}{n}\cdot\frac{1}{n}$$

and it approaches 1/2.

The second one, with the squares, is

$$\left(\frac{1}{n}\right)^2\cdot\frac{1}{n} + \left(\frac{2}{n}\right)^2\cdot\frac{1}{n} + \left(\frac{3}{n}\right)^2\cdot\frac{1}{n} + \cdots + \left(\frac{n-1}{n}\right)^2\cdot\frac{1}{n} + \left(\frac{n}{n}\right)^2\cdot\frac{1}{n}$$

and it approaches 1/3.

What's the next sum after that?

CHRISTOPHER. With cubes,

$$\left(\frac{1}{n}\right)^3\cdot\frac{1}{n} + \left(\frac{2}{n}\right)^3\cdot\frac{1}{n} + \left(\frac{3}{n}\right)^3\cdot\frac{1}{n} + \cdots + \left(\frac{n-1}{n}\right)^3\cdot\frac{1}{n} + \left(\frac{n}{n}\right)^3\cdot\frac{1}{n}$$

approaches 1/4.

SERGE LANG. Who says 1/4? Ann, what do you think?

ANN. Nothing, I don't know.

SERGE LANG. Elisabeth? Siem Lang?

SIEM LANG. 1/4.

SERGE LANG. We proved it for the first two sums. But this third one, we have not proved it. And the next one, what is it going to be?

ELISABETH.

$$\left(\frac{1}{n}\right)^4 \cdot \frac{1}{n} + \left(\frac{2}{n}\right)^4 \cdot \frac{1}{n} + \left(\frac{3}{n}\right)^4 \cdot \frac{1}{n} + \cdots + \left(\frac{n-1}{n}\right)^4 \cdot \frac{1}{n} + \left(\frac{n}{n}\right)^4 \cdot \frac{1}{n}$$

approaches 1/5.

SERGE LANG. Terrific. That's right! It's a theorem. But how are we going to prove it?

PHILIP. There are no dimensions which allow us to expand $\left(\frac{1}{n}\right)^3 \cdot \frac{1}{n}$ to find 1/4.

SERGE LANG. That's very good. You have understood what I did very well. So we may be stuck. Up to three dimensions, I managed by drawing pictures and stairs, but if I want to handle the next case . . .

A STUDENT. Well, you could imagine . . .

SERGE LANG. That's right, you can imagine it! And you should. But it's not quite as safe, one feels less secure. It raises some problems, and actually one has to use another method because our intuition in higher dimensions . . . is more delicate. For the sum with cubes, we would have to work in how many dimensions?

PHILIP. Four.

SERGE LANG. Right. And five for the next. But now, it's not so sure any more that we are doing things right, the intuition may break down. So one has to give algebraic arguments to complement the geometric intuition of higher dimensions. One can do it, but I won't do it today. Anyhow, it's true.

The method which I used today, which is to cut up the ball into cylinders and to take the sum, do you know who used it for the first time? It was Archimedes. Have you heard of Archimedes? Christopher . . . Right, that's the guy who shouts "Eureka" when he jumps naked out of his bathtub. You heard about this? He was very clever. And it was he who invented this kind of method. And I copied Archimedes, approximately. So let me summarize:

Volume of ball of radius 1 is equal to $\frac{4}{3}\pi$.

And the volume of the ball of radius r? Christopher? With dilations?

PHILIP.

The volume of the ball of radius r is equal to $\frac{4}{3}\pi r^3$.

SERGE LANG. Yes, you catch the r^3 by dilations, that's right. You are really very good in this class, you are very much on the ball. I have rarely seen a class which does as well. You are all quite smart.

Do you have any questions? On mathematics, or anything else? [*A hand goes up.*] Yes?

A STUDENT (CAROLINE). To compute the surface of a sphere, couldn't you turn the disc around?

SERGE LANG. Yes, you can. [*Laughing.*] Oh! Oh! Yes, yes, you can but . . . OK, let's try. To find the surface of the sphere by turning the disc around the vertical axis.

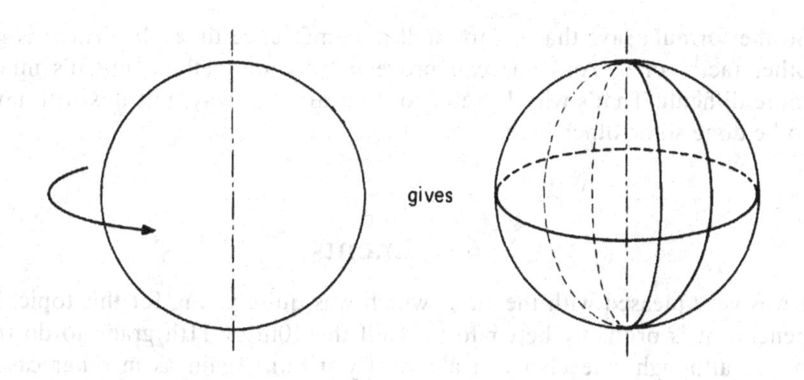

OK, take the circumference, or rather half the circumference, which is πr. Turn it around. Then what happens? You have to multiply πr by something?

THE STUDENT. By the length of the circle in the middle.

SERGE LANG. Yes, but then you will find $2\pi^2 \cdot r^2$.

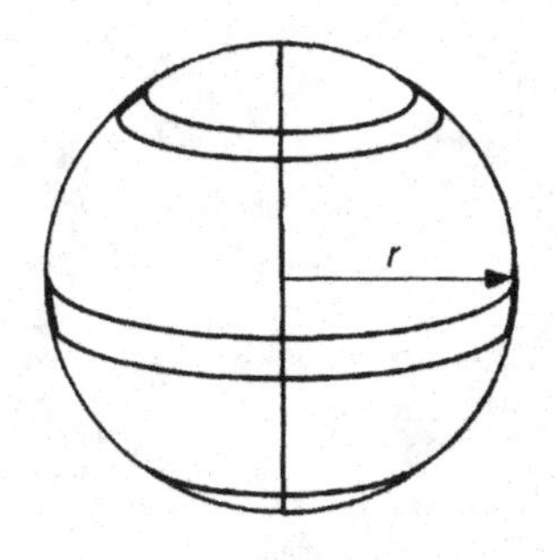

THE STUDENTS. Oh!

SERGE LANG. And that would be true if all the small segments of the half circumference turned around a circle of radius r. But you see, those on top and on the bottom turn around circles which are smaller than that. So the formula $2\pi^2 r^2$ cannot be the right one. I agree that you get the surface of the sphere by turning the half circumference around, but then how do get the right formula? Do you see the problem? There is a difficulty. You have to take account of the fact that certain pieces turn around small circles.

You had a good idea, but you would have to use an "average" circle. There is also a theorem for that, but it's too complicated to do it now. It's not at all clear what happens. Do you know the formula for the area of a sphere? It is that:

$$\boxed{\text{Area of the sphere} = 4\pi r^2.}$$

So the formula says that it's the half circumference times $4r$. There is no other factor of π. And one can prove it by your method, but it's much more difficult. That's why I prefer to do it another way, but this will have to be done some other time.

Comments

I was very pleased with the class, which was quite young for this topic. In general, it is probably better to wait till the 10th or 11th grade to do the proof, although a teacher can always try it out. Again as in other cases, the geometry raises interesting problems, which lead into algebra. I find these topics the perfect setting to practice algebra, and to get students acquainted with π.

The area of the sphere is the subject of the next lectures.

The length of the circle

The next lecture was given to an 8th grade class in Paris in Spring 1983. We did the length of the circle, starting with the formula for the area. This method can be used instead of the method followed in "What is pi?" and has the advantage that it adapts easily to find the area of the sphere from the volume of the ball.

SERGE LANG. So, what do you do in this class?

STUDENTS. [*Laughing.*] Nothing!

SERGE LANG. Oh come on! You must do something! How far have you got?

SOME STUDENTS. Factorizations, identities.

SERGE LANG. And in geometry? We are going to do some geometry today. You know the circle, the disc?

A STUDENT. Yep.

SERGE LANG. You know the surface of a disc?

THE STUDENT. [*Hesitating.*] Yes.

SERGE LANG. What is it?

THE STUDENT. r squared times π.

SERGE LANG. That's right. The area of a disc of radius r is equal to πr^2. And the circumference of the circle?

THE STUDENT. It's π times the diameter.

SERGE LANG. And as a function of the radius?

THE STUDENT. $2\pi r$.

SERGE LANG. Yes. And did you do the proof? How do you prove these two formulas? [*Questioning silence.*]

STUDENTS. Euh . . .

SERGE LANG. Well, that's what I want to do, show you how you can prove the second formula, starting from the first. In other words, let's suppose that we know the first formula, and we want to prove the second. Let me write this down.

We suppose known that the area of the disc is πr^2, and we want to prove the second formula, that

$$c = 2\pi r,$$

where c is the circumference of the circle.

So I have a circle of radius r, and I also draw a circle with a slightly bigger radius, r plus something.

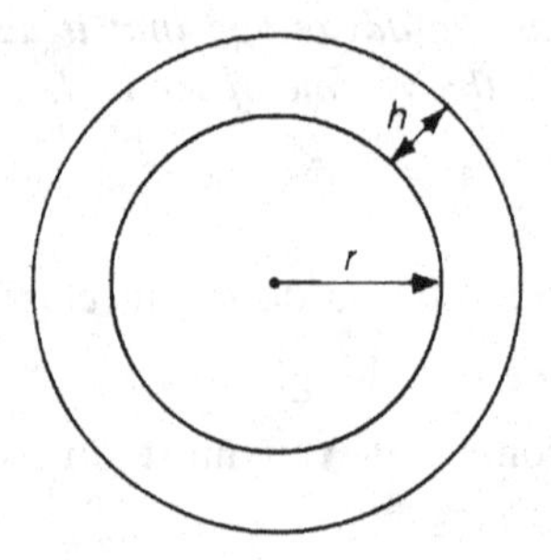

Let's call this something h. Then the radius of the bigger circle is $r+h$.

In between the two circles, I have a band. You see the band, I draw a shading of the band like this.

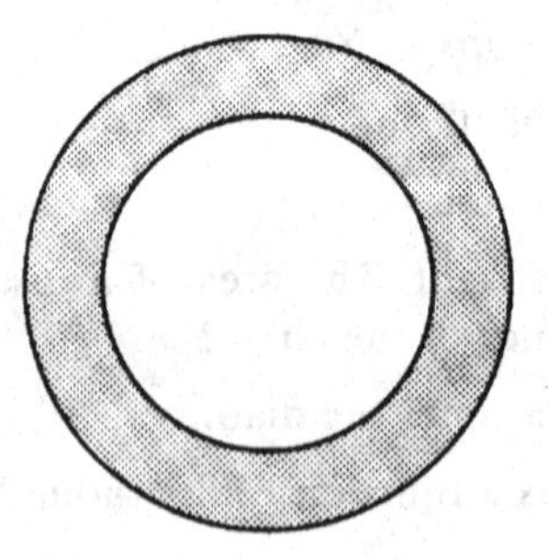

What's the area of the band?

THE CLASS. It's the area of the big disc minus the area of the small disc.

SERGE LANG. Precisely, the area of the big disc minus the area of the small disc. But we know these areas, and I can write down the formulas for them. What's the area of the small disc?

A STUDENT. It's πr^2.

SERGE LANG. Right. And the area of the big disc? It's equal to what? [*Questioning looks.*]

We know the formula for a disc of radius r, it's πr^2. If the disc has radius 2, what is its area? It's $\pi 2^2$, so 4π. OK? Now the area of a disc of radius 3 is equal to what? You, what's your name?

THE STUDENT. Nancy. It's π times 3 squared, so it's 9π.

SERGE LANG. Right. And the area of a disc of radius 4, what is it? Nancy.

NANCY. $\pi 4^2$, so 16π.

SERGE LANG. OK, and the area of a disc of radius 10?

NANCY. $\pi 10^2$.

SERGE LANG. And the area of a disc of radius r?

NANCY. πr^2.

SERGE LANG. And the area of a disc of radius $r+h$?

NANCY. [*Hesitating*] $\pi(r+h)^2$?

SERGE LANG. Yes, you got it, $\pi(r+h)^2$. Now we can write the area of the band. It's equal to

$$\pi(r+h)^2 - \pi r^2.$$

Do you all agree? [*The students approve.*] Now what do we do to this formula?

A STUDENT. We can work it out.

SERGE LANG. Yes, and then $(r+h)^2$, what do we do with it? [*Turning toward the teacher:*] You have done this?

TEACHER. I should say so!

SERGE LANG. $(a+b)^2$ is equal to what?

STUDENTS, ALL TOGETHER: $a^2 + 2ab + \cdots$

SERGE LANG. Wait, not all together. You, what's your name?

STUDENT. Rita.

SERGE LANG. OK, Rita, $(a+b)^2$ is equal to what? . . .

A STUDENT. It's an algebraic identity.

SERGE LANG. Who knows?

A BOY. It's $a^2 + 2ab + b^2$.

SERGE LANG. That's right, $(a+b)^2 = a^2 + 2ab + b^2$. You must have done this before, no?

STUDENTS. We did it with v and w.

SERGE LANG. When I was a kid, we did it with a and b, but it's the same thing. And it's again the same thing if we do it with r and h. So $(r+h)^2$ is equal to what?

STUDENTS. [*All together*]: $r^2 + 2rh + h^2$.

SERGE LANG. Very good, $r^2 + 2rh + h^2$. And how do you prove it?

STUDENTS. Take the product.

SERGE LANG. Yes, let's do it in detail. $(r+h)^2$ it's just $(r+h)$ times $(r+h)$. Now what do I do? [*All students talk at once. Serge Lang writes:*]

$$(r+h)(r+h) = r(r+h) + h(r+h).$$

And after that, what do we do? You, what's your name?

A BOY. Christopher. We get $r^2 + rh + hr + h^2$.

SERGE LANG. So . . .

CHRISTOPHER. $r^2 + 2rh + h^2$.

SERGE LANG. You got it. So we have proved the formula for $(r+h)^2$. Now to learn it by heart, we do it with a and b, or v and w, or what? Let's take a and b. Then a plus b squared is equal to a squared plus two ab plus b squared. Let's repeat this all together.

SERGE LANG AND THE CLASS. a plus b squared equals a squared plus two ab plus b squared. a plus b squared equals a squared plus two ab plus b squared.

SERGE LANG. OK? You got it in your ears? You'll never forget it? That's the way one memorizes something, you learn it by heart. You repeat it three or four times at night, before going to bed, and the next day, you know it. That's the way I learned it.

So with $(r+h)^2$?

STUDENTS. $r+h$ squared equals r squared plus two rh plus h^2.

SERGE LANG. Good. Now we can do the subtraction:

$$\pi(r+h)^2 - \pi r^2.$$

It's equal to what? It's equal to

$$\pi(r^2 + 2rh + h^2) - \pi r^2.$$

So what do we get. You [*pointing to a student*], what's your name?

A STUDENT. Jerry.

SERGE LANG. So what do you get?

JERRY. $2rh + h^2$.

SERGE LANG. Is that true? [*Some students indicate something is missing.*] Watch out! We had a π in front of the parenthesis. Then we get πr^2

and $-\pi r^2$, which cancel, and we still have left π times $2rh + h^2$. Right? Then we get a formula which gives us the area of the band. I write it down:

The area of the band is equal to
area of the disc of radius $(r+h)$ − area of disc of radius r
in other words
area of the band $= \pi(2rh + h^2)$.

OK? No protest?

A STUDENT. No, we trust you. [*Laughter.*]

SERGE LANG. No, no! You should not trust me! The question is whether you have understood or not.

A STUDENT. So so.

SERGE LANG. Did you understand how we got the area of the band?

THE STUDENT. Heu . . . [*Looking at the picture.*] We took the big circle minus the small one.

SERGE LANG. That's almost it. We took the area of the big circle minus the area of the small circle. And the area of a disc of radius r is πr^2. Then the radius of the big disc is equal to what?

CHRISTOPHER. $r+h$.

SERGE LANG. And the area? It's $\pi(r+h)^2$. After that, I subtract the area of the disc of radius r, which is r^2. Then we do a little algebra to compute $(r+h)^2$, and we find the expression

$$\text{area of band} = \pi(2rh + h^2).$$

Are you convinced?

CHRISTOPHER. Yep.

SERGE LANG. By conviction or intimidation?

CHRISTOPHER. By conviction.

SERGE LANG. Good, so now we have found the area of the band. Next suppose I have a rectangle, like this:

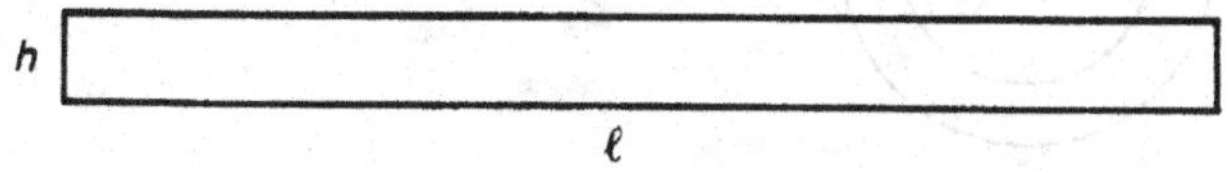

h ℓ

with base of length l, and height h. Then the area of the rectangle is what? [*Pointing to a student.*]

THE STUDENT. . . .

SERGE LANG. The area of a rectangle is equal to what?

THE STUDENT. It's the base times the height.

SERGE LANG. That's right, it's lh. You do know this or not?

CHRISTOPHER. Yes, we know it.

SERGE LANG. Good. Then what I want to do, is to find another expression for the area of the band. I want to find inequalities for the area of the band. A while ago, I found an exact expression. Now I want to find another way of expressing the area of the band. Approximately, it will be the length times the height.

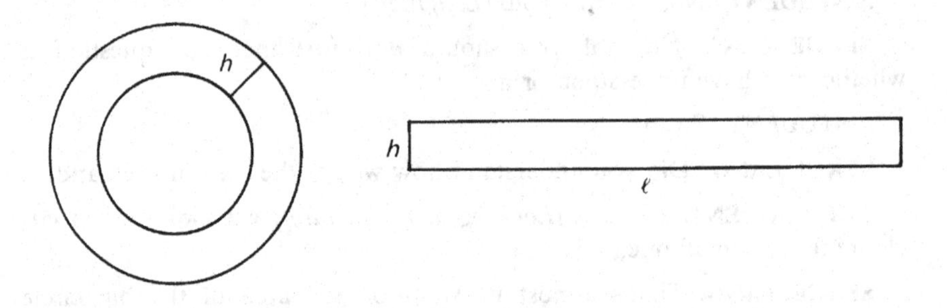

But I have two possible lengths, with the two circles. I have the length of the small circle, and I have the length of the big circle. There is a small circumference and a big one. And the area of the band will satisfy certain inequalities. It is bigger then the small circumference times the width. Who sees this? Who sees that the area of the band is bigger then the length of the small circle times the width of the band? [*Serge Lang points to several students.*] You see this.

THE STUDENT. Yes, it's OK.

SERGE LANG. It's like on this picture. I fold the band over the circle.

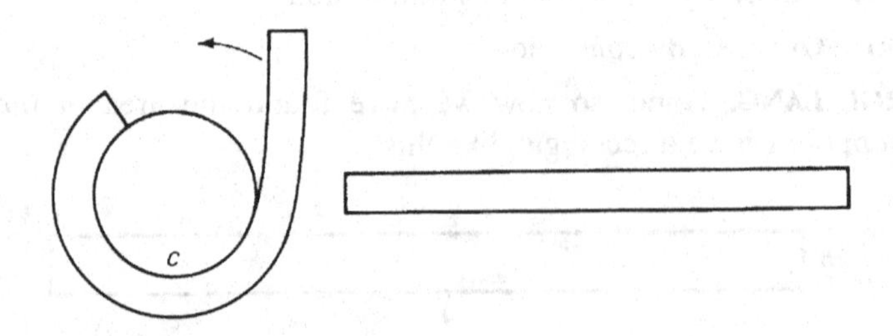

So I find the inequality:

(small circumference) × height ≦ area of the band.

A STUDENT. I have not quite understood why.

SERGE LANG. Suppose that the length of the rectangle is equal to the small circumference. If I curve the rectangle and wind it around the small circle, then the rectangle sticks exactly around the small circle, but I have to stretch it so it sticks around the bigger one. And if I have to stretch it, this means that the area of the rectangle is smaller than the area of the band. You see this?

THE STUDENT. Yes, I see it now.

SERGE LANG. Good. But on the other side, what kind of an inequality do I get?

A STUDENT. [*Hesitating.*] It's going to be bigger than . . . heu . . . smaller . . .

SERGE LANG. Yes, you are on the right track. I have the band, I have the big circle, and I have the distance between the two circles, which I call the height h. I take a rectangle whose base is equal to the big circumference, the circumference of the big circle, and whose height is h. Now I curve this rectangle, to get a curved band. The area of the rectangle is simply the product of the big circumference times h. And the area of the rectangle is bigger then the area of the curved band, because if I curve the rectangle so it sticks exactly along the big circle, then it's going to fold up along the small circle, it's going to make pleats.

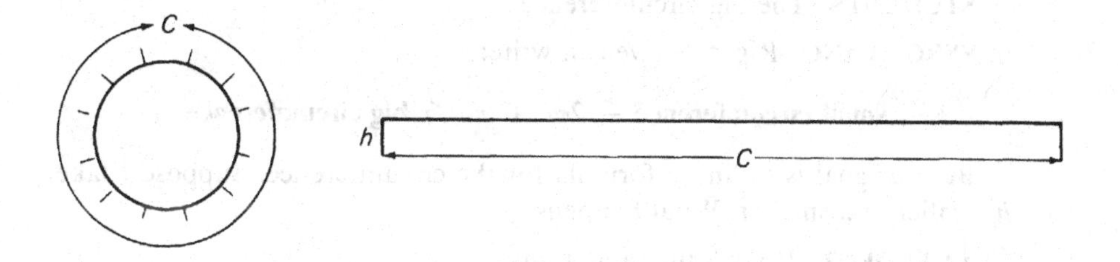

Do you see it? Then I can write the two inequalities:

(small circumference) $h \leqq$ **area of the band** $\leqq$ **(big circumference)** h.

Now, let's divide everything by h. I divide by h on the left, on the right, and in the middle. What do I get?

[*Questioning silence in the class.*]

SERGE LANG. If you know that $a = b$, then do you know that $ah = bh$?

THE CLASS. Yes.

SERGE LANG. And conversely, if $ah = bh$ and h is not zero, then what?

THE CLASS. Then $a = b$.

SERGE LANG. Now, if $a \leqq b$ then $ah \leqq bh$ and if $ah \leqq bh$ then . . . ? If h is positive?

A STUDENT. $a \leqq b$.

SERGE LANG. Right. If h is positive. So if I have something which is smaller than another thing, and if I divide both things by h, then I still have this same inequality. So what do we get? When I divide on the left by h, then I get the circumference. When I divide the middle by h, I get the area of the band divided by h. Which gives what? [*Pointing to a student.*]

[*The student keeps quiet.*]

SERGE LANG. What's the area of the band? We found the formula a minute ago.

THE STUDENT. $\pi(2rh + h^2)$.

SERGE LANG. And if I divide by h?

STUDENT. $2\pi r$.

SERGE LANG. Yes, that's the first term. And the second term?

SEVERAL STUDENTS TOGETHER. πh.

SERGE LANG. Very good. So we find $2\pi r + \pi h$. Now, on the right, what do we get?

STUDENTS. The big circumference.

SERGE LANG. Right. So we can write:

$$\textbf{small circumference} \leqq 2\pi r + \pi h \leqq \textbf{big circumference.}$$

But our goal is to find a formula for the circumference. Suppose I take h smaller and smaller. What happens?

A STUDENT. It gives the same thing.

SERGE LANG. It gives the same thing, yes. In other words, the big circumference approaches the small circumference. So if h approaches zero, what happens? The small circumference stays whatever it is, it does not change. The middle term is $2\pi r + \pi h$. Then $2\pi r$ does not change, it does not depend on h. But πh, what happens to πh when h becomes smaller and smaller?

A STUDENT. It becomes different.

SERGE LANG. It becomes different, OK. But how? Well, π is approximately equal to 3.14. If I take $h = 1/10$ then πh is approximately equal to what?

CHRISTOPHER. 0.314.

SERGE LANG. And if $h = 1/100$?

SEVERAL STUDENTS. 0.0314.

SERGE LANG. And if $h = 1/1000$?

THE CLASS. 0.00314.

SERGE LANG. Now if h becomes smaller and smaller, then what happens to πh?

THE CLASS. It becomes smaller and smaller.

SERGE LANG. That's right, and therefore it approaches zero. And $2\pi r + \pi h$ approaches what?

THE CLASS. The small circumference.

SERGE LANG. You are right, $2\pi r + \pi h$ approaches the small circumference, but when h approaches zero, then we have $2\pi r$ plus a number which becomes smaller and smaller:

if $h = 1/10$ then we have $2\pi r + 0.314$;
if $h = 1/100$ then we have $2\pi r + 0.0314$;

after that, we get $2\pi r + 0.00314$; then $2\pi r + 0.000314$. And if I continue like this, when h becomes smaller and smaller, the sum approaches what?

A STUDENT. It approaches $2\pi r$.

SERGE LANG. Yes. Now on the right hand side, I have the big circumference. When h approaches zero, what does the big circumference approach? You already said it.

THE CLASS. [*All together*] The small circumference.

SERGE LANG. Very good. So when h approaches zero, then I get the small circumference on each side, on the left and on the right, and I have $2\pi r$ in the middle. This means that $2\pi r$ is squeezed between these two numbers: one is the small circumference c, and the other is the big circumference C. When the width h approaches zero, then C approaches c, and since we squeezed $2\pi r$ between them, we get

$$c = 2\pi r.$$

That's just what we wanted to prove!

Remember, we started from the area of a disc, and we found the formula for the circumference.

Now if we go to dimension 3, it won't be a disc any more, it will be what?

A STUDENT. A sphere.

SERGE LANG. Yes, you are right. We can draw a sphere, and another slightly bigger. I don't have the time to do this now, but suppose I know the volume of the ball. What do I do to find the surface of the sphere. Christopher, do you know?

CHRISTOPHER. . . .

SERGE LANG. What did we do for the circle. Do you remember?

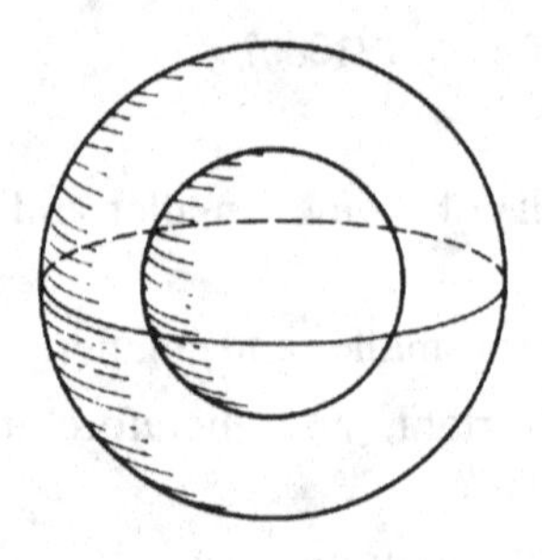

CHRISTOPHER. No.

SERGE LANG. We just did it. Let's try to go through the proof quickly. We started with the area of the disc, and we computed the area of the band. How did we do it? [*Pointing*] You, what's your name?

STUDENT. Roger.

SERGE LANG. OK, Roger, how did we compute the area of the band?

ROGER. We took the difference between the two discs.

SERGE LANG. OK. And after that, we wrote inequalities for the area of the band. You, what's your name?

STUDENT. Yaelle.

SERGE LANG. So then, Yaelle?

YAELLE. We said that the area of the band is bigger then the product of the small circumference times h.

SERGE LANG. And after that, what did we do?

YAELLE. You did the same thing with the big circumference.

SERGE LANG. Yes. And then?

YAELLE. You cancelled h and divided by h.

SERGE LANG. Right. Now when h becomes smaller and smaller, the small circumference remains whatever it is, and the big one?

YAELLE. It approaches the small one?

SERGE LANG. Now you have to know what happens to the expression in the middle, which is the area of the band. Before that, we still had one thing to do. Roger?

ROGER. You took the area of the big disc minus the area of the small disc.

SERGE LANG. Right, and that was $\pi(r+h) - \pi r^2$. And after that, what did we do?

ROGER. You used an identity.

SERGE LANG. Yes, and what did we find?

ROGER. $2\pi rh + \pi h^2$.

SERGE LANG. Very good. And Yaelle said that we divided everything by h. Then what do we get?

ROGER. $2\pi r + \pi h$.

SERGE LANG. Now, suppose that h becomes smaller and smaller. What happens to πh? Let's pick on somebody else. You! [*Pointing.*]

THE STUDENT. It becomes smaller and smaller.

SERGE LANG. That's right.

When h approaches 0, then πh also approaches 0.

And the big circumference . . . ?

THE STUDENT. It approaches the small one.

SERGE LANG. Right, the big circumference approaches c, which is the small circumference. Therefore, we find:

$$c \leqq 2\pi r \leqq c,$$

and this means that $c = 2\pi r$. And that's the proof.

OK, who can repeat the entire proof?

[*Here, Serge Lang asks another student to repeat the proof, and the student still needs to be guided through it.*]

OK, let's stop this for today, and let's talk about something else.

A STUDENT. It's complicated.

SERGE LANG. It was too hard? We should go through it one more time.

THE CLASS. [*All together.*] No.

SERGE LANG. No, not today, not today.

A STUDENT. But we are not used to this.

SERGE LANG. Oh, it's not that bad. What did we need? Just $(r+h)^2$. It's not that complicated.

THE STUDENT. No.

[*General discussion among the students.*]

SERGE LANG. Do you do proofs like this, in class?

THE STUDENT. Not like this one.

SERGE LANG. You want to talk about something else?

A STUDENT. What do you do? What's your work?

SERGE LANG. I am a professor of mathematics.

STUDENT. And physics?

SERGE LANG. No, just mathematics, at the university. I don't know much physics. And I do research.

A STUDENT. [*Disbelieving.*] In mathematics?

SERGE LANG. The theorem that we just proved, somebody had to prove it for the first time once.

A STUDENT. Oh yes.

SERGE LANG. It was done long ago, but there are other things which have not yet been discovered.

YAELLE. And you look for them?

SERGE LANG. Yes.

ANOTHER STUDENT. What if you don't find anything? [*Laughter.*]

SOME STUDENTS. Resign! Resign! [*Laughter.*]

SERGE LANG. If I don't find anything? Yes, it can happen, but if I did not find anything, ever, then I would not be a research mathematician. I would not have taken this direction, and I would have done something else.

A STUDENT. You find theorems?

SERGE LANG. Yes, some.

STUDENT. About what?

SERGE LANG. Oh you know, it's at a level which I can't explain here, it would take too much preparation.

THE STUDENT. People will learn them in 200 years?

SERGE LANG. In high school? Maybe, but there are people who learn them now.

[*The discussion continued on a variety of topics, including "Dallas", which the French kids watch just like the American ones. One kid said that her French teacher called "Dallas" "feeble minded". There was no disagreement, but they watch it. The students also brought up American culture and its influence on French culture, and economic problems faced by their country. It was remarkable on the whole to see how television influences their thinking and makes them aware of contemporary problems. After nearly an hour, I felt we should not end that way, but that we should go back to mathematics. So I picked up the reins once more amidst the general brouhaha.*]

SERGE LANG. Look, the time is almost over, we can't end like this. Let's finish with some mathematics. Maybe we can go through the proof once more.

A FEW STUDENTS. Not again!

SERGE LANG. Let's come back to the proof. [*Pointing to Yaelle:*] You! I started from the area of the disc πr^2, and wanted to find $2\pi r$ for the circumference. So how did we do it?

[*The brouhaha stops as soon as Yaelle starts to speak. All the students remain quiet and listen intently as she goes through the proof.*]

YAELLE. Well, you took the area of the big disc which is of radius $r+h$, and you subtracted the area of the small disc which has radius r . . .

SERGE LANG. [*Encouragingly*] Yes.

YAELLE. So this gives, after you work out the product: $2\pi rh + \pi h^2$. Then you had some inequalities, and you divided by h . . .

SERGE LANG. Hold it! Which inequalities?

YAELLE. The area of the band is bigger than the circumference of the small circle times h, and it is smaller than the circumference of the big circle times h.

SERGE LANG. Right.

YAELLE. Then, you cancelled the two h's dividing the area of the band by h. This gave that when it approached . . . euh . . . wait . It gives $2\pi r$.

[*She hesitates.*]

SERGE LANG. Yes, you are doing OK. You get $2\pi r + \pi h^2$ in the middle; and on the two sides?

YAELLE. On the left, you have the small circumference, and on the right you have the big circumference, and we saw that when h becomes smaller and smaller, it approaches c, and we concluded that c was smaller than $2\pi r$ and also that $2\pi r$ was smaller than c, so it gives $2\pi r$ equals c.

SERGE LANG. [*Very happy with this performance—the students also*] That's right! Congratulations. You *have* learned something.

[*The entire class applauds Yaelle warmly.*]

SERGE LANG. It's really good. I am very happy to have met all of you. It went very well . . . Yes, it was really good.

Comments

I was talking to quite a young class (8th grade). Shortly after, I gave a similar talk to a 10th grade class (unfortunately not taped), and certain difficulties which I encountered with the younger students had disappeared for this more advanced group. In both cases, I asked a student at the end to repeat the proof, and in both cases, it was a success, giving rise to the spontaneous applause. This shows how much students like to learn and understand mathematics.

More technically, we see here how we can insert technical considerations (like the identity for $(a+b)^2$) in a conceptual context. Starting with a problem which students grant is interesting (finding the length of the circle), we meet a technical question, which is solved as we go along. We mix the general proof, based on general ideas, with such technical material, and with rote memorization achieved by repeating the formula $(a+b)^2 = a^2+2ab+b^2$ out loud. The formula must be learned as a conditioned reflex, and students should be able to reproduce it instantaneously. This is complementary to being able to reproduce the proof for the main theorem.

The area of the sphere

I had the previous class only for an hour, and so was able to do only the length of the circle with them. But Stephane Brette (14 years old) was visiting from another school, and so we were able to do the area of the sphere together right after the class, by the same method.

SERGE LANG. [*Addressing Stephane.*] So, you came from another high school to listen to today's class. Did you understand what I did?

STEPHANE. Yes.

SERGE LANG. Could you do the same thing to find the area of a sphere? Suppose I give you the volume of the ball,

$$V = \frac{4}{3}\pi r^3.$$

What's the area of a sphere of radius r?

STEPHANE. I forgot the formula.

SERGE LANG. The formula is

$$A = 4\pi r^2.$$

OK, so you don't forget it again, we are going to repeat it together. [*And we repeat together out loud: the area of a sphere is* $4\pi r^2$, *the area of a sphere is* $4\pi r^2$ . . .].

You won't forget it any more?

STEPHANE. No.

SERGE LANG. Good. Then go to the blackboard, and prove the formula.

[*Stephane goes to the blackboard and draws the following picture.*]

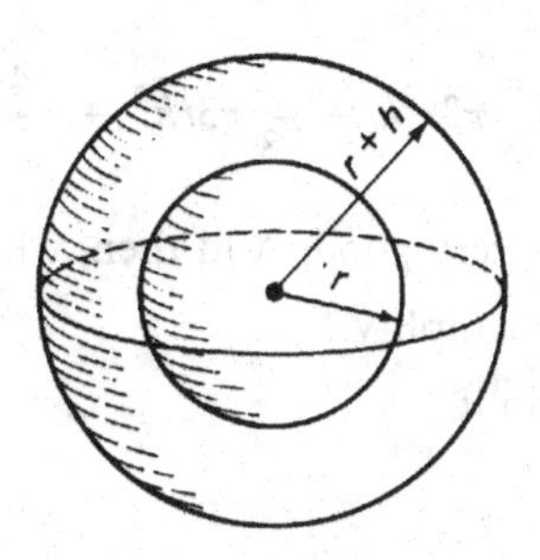

SERGE LANG. That's right, you take two balls, one of radius r and the other of slightly bigger radius $r+h$. And then?

STEPHANE. Then you have inequalities. [*And Stephane writes:*]

(area of the small sphere) h $\leqq$ volume of the band
$\leqq$ (area of the big sphere) h.

SERGE LANG. Yes, you have a band between the two spheres, and these are the inequalities which correspond to the band between two circles, but now we are in dimension 3. The volume of the band is equal to what?

STEPHANE. It's the volume of the big ball minus the volume of the small ball.

SERGE LANG. OK, so write them down.

[*Stephane writes:*]

Volume of the band

$$= \text{volume of the big ball} - \text{volume of the small ball.}$$

$$= \frac{4}{3}\pi(r+h)^3 - \frac{4}{3}\pi r^2.$$

SERGE LANG. There you are. But you see, instead of $(r+h)^2$ and r^2 for the circle, we now have $(r+h)^3$ and r^3 for the ball. We need some identity for the expansion of $(r+h)^3$. Do you know it?

STEPHANE. No, we haven't done that in class.

SERGE LANG. It doesn't matter, I'm going to show you the answer now, you can work with it, and then we'll prove it afterwards. The identity states that

$$(r+h)^3 = r^3 + 3r^2h + 3rh^2 + h^3.$$

Now substitute. What do you find?

STEPHANE. [*Writes on the blackboard:*]

$$\frac{4}{3}\pi(r^3 + 3r^2h + 2rh^2 + h^3) - \frac{4}{3}\pi r^3$$

$$= \frac{4}{3}\pi r^3 + \frac{4}{3}\pi 3r^2h + \frac{4}{3}\pi 3rh^2 + \frac{4}{3}\pi h^3 - \frac{4}{3}\pi r^3.$$

SERGE LANG. That's very good. And then, what happens?

STEPHANE. You can simplify.

SERGE LANG. OK, do it.

STEPHANE. One can simplify $\frac{4}{3}\pi r^3$. Then we get

$$4\pi r^2 h + 4\pi r h^2 + \frac{4}{3}\pi h^3.$$

SERGE LANG. And then?

STEPHANE. Well, after, you divide everywhere by h.

SERGE LANG. So go ahead, divide by h. What do you find?

STEPHANE. [*Writes:*]

area of small sphere $\leqq 4\pi r^2 + 4\pi r h + \frac{4}{3}\pi h^2 \leqq$ area of big sphere.

SERGE LANG. Now, what do we do?

STEPHANE. The big sphere approaches the little sphere, and the area of the big sphere approaches the area of the little sphere.

SERGE LANG. Precisely. And the expression in the middle is squeezed between the two. What happens to the expression in the middle? For instance, the term $4\pi r h$ approaches what when h becomes smaller and smaller?

STEPHANE. It approaches zero.

SERGE LANG. Yes, and the other term also approaches zero, the $\frac{4}{3}\pi h^2$. So the middle expression approaches what when h approaches zero?

STEPHANE. It approaches $4\pi r^2$.

SERGE LANG. There you are! So if A is the area of the sphere of radius r, then we find

$$A \leqq 4\pi r^2 \leqq A,$$

and therefore we get the equality

$$A = 4\pi r^2.$$

STEPHANE. That's what we were supposed to prove.

SERGE LANG. You said it, that's what we were supposed to prove. You see that the proof went just as easily as the proof for the length of the circle.

All that remains to do is to deal with the little identity for $(r+h)^3$. How do we take care of that?

STEPHANE. You multiply.

SERGE LANG. Yes, we multiply like this [*and Serge Lang writes on the blackboard*:]

$$\begin{aligned}(r+h)^3 &= (r+h)(r+h)^2 = (r+h)(r^2 + 2rh + h^2)\\ &= r(r^2 + 2rh + h^2) + h(r^2 + 2rh + h^2)\\ &= r^3 + 2r^2h + rh^2 + hr^2 + 2rh^2 + h^3\\ &= r^3 + 3r^2h + 3rh^2 + h^3,\end{aligned}$$

and we have proved the identity.

Now that we have given the proof, it's worth while to learn the formula by heart, just like $(r+h)^2$. There are similar formulas for $(r+h)^4$, $(r+h)^5$ and so on, but they are a little more complicated. You can experiment with these higher expressions, one after the other, to see what goes on, and to see if you can find the general rule. But that's enough for today. You can do this some other time.

Pythagorean triples

The following talk is the first of two given to an 11th grade class in a high school in the suburbs of Toronto in Spring 1982. The students were about 16 years old. The talk lasted about 50 minutes.

SERGE LANG. What is this class? Eleventh grade? [*Students nod.*] And what do you guys do? Algebra?

A STUDENT. Nothing. [*Laughter. Many students talk at once.*]

SERGE LANG. Well, you must do *something*. [*Laughter.*] I'm going to do mostly algebra, but I start from a problem with geometric overtones. This problem goes back to Euclid. Suppose you have a right triangle, with sides a, b, c.

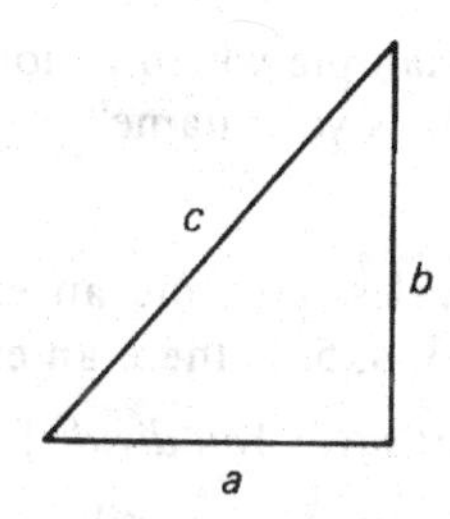

You guys know the Pythagoras theorem?

A STUDENT. Yeah.

SERGE LANG. What's Pythagoras' theorem?

A STUDENT. $a^2 + b^2 = c^2$.

SERGE LANG. That's right. Now can you give me an example of a right triangle where a, b, c, are integers? You know what an integer is?

[*Students give answers. One of them answers the question about a right triangle:*]

A STUDENT. 3, 4, 5.

SERGE LANG. Yes, because $3^2 + 4^2 = 5^2$; $9 + 16 = 25$.

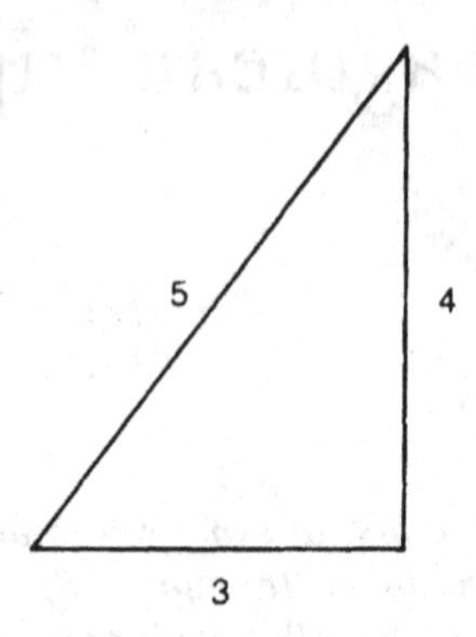

Can you give me another one?

A STUDENT. 6, 8, 10. [*Laugher.*]

SERGE LANG. That's not so stupid. 6, 8, 10 is another one. How did you get 6, 8, 10?

THE STUDENT. I multiplied 3, 4, 5 by 2.

SERGE LANG. That's right. But you could also multiply them by 3, right? You can multiply them by anything, by any integer (a whole number). For example, if d is any integer, then

$$(3d)^2 + (4d)^2 = 3^2d^2 + 4^2d^2 = (3^2 + 4^2)d^2 = 5^2d^2 = (5d)^2.$$

Now can you give me an example which is not obtained by multiplying 3, 4, 5 by an integer? You, what's your name?

STUDENT. Charles.

SERGE LANG. OK, Charles, give me an example with integers a, b, c which are not multiples of 3, 4, 5. Is there an example like that?

[*Long silence. One student says: You divide.*]

SERGE LANG. If you divide, do you still get an integer?

STUDENT. No.

SERGE LANG. But I want integers.

ANOTHER STUDENT. 5, 12, 13.

SERGE LANG. Let's see, $5^2 = 25$, $12^2 = 144$ and $25 + 144 = 169$, which is 13^2. It works. All right. Can you give me another one?

[*Silence.*]

Think hard. Is there another one?

A STUDENT. 10, 24, 26.

SERGE LANG. Yes, but that's a multiple of 5, 12, 13. You multiplied them by 2. You see, once you have a solution, you can take any multiple to get another one. Try to give me another one which is not a multiple of the solutions we already have. I don't want multiples.

A STUDENT. 7, 24, 25.

SERGE LANG. Let's see. It seems to work. How did you get it, by the way?

STUDENT. [*The answer cannot be understood on the tape.*]

SERGE LANG. Well, obviously you compute pretty fast. That's very good. Let's check it out. Somebody check it out quick.

SOME STUDENTS. It's OK.

SERGE LANG. Can you give me another one? [*Silence.*] I mean, look. If you try to compute one by one like that . . .

A STUDENT. I can find another one.

SERGE LANG. Wait a minute. Suppose you try to compute like that. Sure, maybe you can find another one. The next question is, will you be able to find infinitely many? [*Silence.*] Do you understand the question?

STUDENTS. Yes.

SERGE LANG. So how many say yes, there are infinitely many? Will you be able to find infinitely many which are not obtained by multiplying one of them by an integer, which are not multiples of a solution? How many say you can?

[*Some hands rise.*]

How many say you cannot?

[*Many hesitate, some hands rise.*]

How many don't know? How many keep a prudent silence? [*Laughter.*]

SERGE LANG. [*Pointing to a student.*] What do you think?

THE STUDENT. Infinitely many.

SERGE LANG. Why do you think that?

THE STUDENT. Right now you can tell.

SERGE LANG. Oh no, you can't tell. Right now you have given me three.

STUDENT. I don't know.

ANOTHER STUDENT. There are infinitely many squares. They go on forever.

SERGE LANG. Yes, the squares go on forever. There are infinitely many squares. But I want to find squares which are related by this equation. That's the question. Do *they* go on forever?

STUDENT. Yeah, because eh . . . you had a first one, 3, 4, 5. Then you have 5, 12, 13. Then 7, 24, 25. So the first numbers in each thing are 3, 5, 7. Then the next one would be 9, but that might be . . . I would get a multiple of 3.

SERGE LANG. Yes, this might give a multiple of 3. And then what? Do you think you might find one with 11?

STUDENT. I am not sure if you can.

SERGE LANG. It's very nice what he is doing. He is really thinking very well.[1] And that, indeed, is how you might go about trying to determine whether there are infinitely many or not. Now what I am going to do, is first to show you that there are infinitely many, and then to describe them all. Absolutely all. That's what we are going to do. How many think it's an interesting question?

[*Laughter.*]

SERGE LANG. Why did you laugh? [*More laughter.*] [*Several students say they find it interesting.*]

SERGE LANG. And are there people who don't find it interesting? [*Laughter.*] Ah, there are a number of hands up, some find it interesting, and others don't find it interesting. Well, OK, so I do it for those who find it interesting. OK?

So I want to find all solutions in integers a, b, c for the equation

$$(*) \qquad a^2 + b^2 = c^2.$$

Suppose I divide the equation (*) by c^2. I get

$$\left(\frac{a}{c}\right)^2 + \left(\frac{b}{c}\right)^2 = 1.$$

OK? Now if a and c are integers, and b, c are integers, what do we call a/c and b/c ? Do you know these numbers are called?

A STUDENT. Rational numbers.

SERGE LANG. Yes. So I let

$$x = a/c \qquad \text{and} \qquad y = b/c.$$

Then x, y are rational numbers. So what I have obtained is the equation

$$(**) \qquad x^2 + y^2 = 1.$$

The equation $x^2 + y^2 = 1$ is the equation for what?

A STUDENT. A circle.

---

[1] Ideally, I should now continue this exchange and let the student go on with his correct line of discovery. I should involve others in a similar activity, and ask them to think about it until the next class. Then I would pursue the topic in the next class after they have had a chance to show their own originality. But I am pressed for time, in this isolated lecture, so I choose to go on.

SERGE LANG. Yes, it's the equation of a circle.

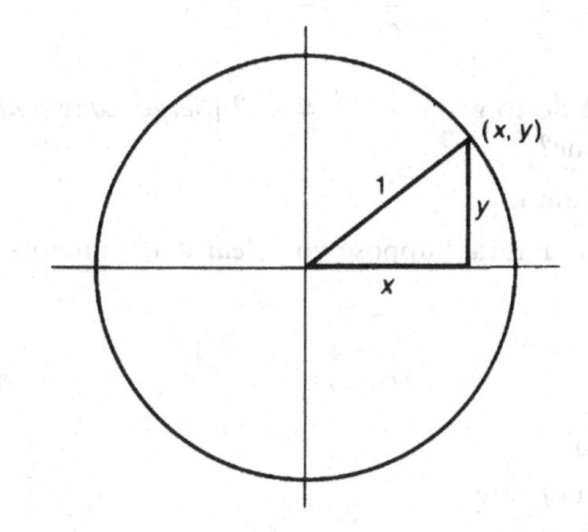

What's the radius of the circle?

STUDENT. One.

SERGE LANG. So I have a circle of radius 1, centered at the origin. Here I have a point (x, y), and what I want to find are all the rational points on the circle. That means all the points with coordinates x, y which are rational numbers, and which satisfy the equation $x^2 + y^2 = 1$. If I have any such rational point, if I have rational numbers x and y such that $x^2 + y^2 = 1$, how do I get a solution of $a^2 + b^2 = c^2$, in integers?

STUDENT. I don't know.

SERGE LANG. What's a rational number?

[*Silence.*]

SERGE LANG. It's a fraction, it's a quotient of two integers, right? Suppose I have two fractions x and y such that $x^2 + y^2 = 1$. How do I find integers a, b, c such that $a^2 + b^2 = c^2$? You can write x and y as fractions with a certain denominator. Any fraction can be written as a numerator over a denominator. Suppose I have two fractions x and y. You can write them over a common denominator. You know that? So let c be a common denominator, so you can write x as a numerator over c; and y as a numerator over c also. So you can write

$$x = \frac{a}{c} \quad \text{and} \quad y = \frac{b}{c},$$

with a and b integers. OK? If I have such fractions, then what will be the relation between a, b and c? If $x^2 + y^2 = 1$ then I have

$$\left(\frac{a}{c}\right)^2 + \left(\frac{b}{c}\right)^2 = 1,$$

which I can also write as

$$\frac{a^2}{c^2} + \frac{b^2}{c^2} = 1.$$

Now what can I do to get $a^2 + b^2 = c^2$? [*Serge Lang points to a student.*] What's your name?

STUDENT. Laura.

SERGE LANG. Laura. Suppose you clear denominators in this equation

$$\frac{a^2}{c^2} + \frac{b^2}{c^2} = 1.$$

What do you get?

LAURA. I'm not sure.

SERGE LANG. What does it mean to clear denominators?

ANOTHER STUDENT. [*He starts giving the right answer.*]

SERGE LANG. You shut up! [*Laughter.*] I know *you* can answer. The point is not with you, the point is with her. All right?

LAURA. You multiply by c^2.

SERGE LANG. That's right. You multiply by c^2. And what do you get when you multiply by c^2?

LAURA. You get $a^2 + b^2 = c^2$.

SERGE LANG. That's right. You get $a^2 + b^2 = c^2$.

If I started with $a^2 + b^2 = c^2$ then I ended up with $x^2 + y^2 = 1$. If a, b, c were integers, x and y are rational numbers. If I start with rational numbers, with fractions x and y, and put them over a common denominator, and then clear the denominators, I end up with $a^2 + b^2 = c^2$. So whether I work with the equation $x^2 + y^2 = 1$ with rational numbers, or with the equation $a^2 + b^2 = c^2$ with integers, I have equivalent problems. Right?

OK, so let's look at $x^2 + y^2 = 1$. We have the problem of getting solutions to the equation. To find all rational numbers x and y such that $x^2 + y^2 = 1$. Now we'll follow this idea how you can get them. I'm going to write down some formulas. Suppose you have played with this idea a lot, how you can get them. Then you would end up with the following formulas. Let

$$x = \frac{1-t^2}{1+t^2}, \qquad y = \frac{2t}{1+t^2}.$$

I let x and y be those expressions. OK? Then what is $x^2 + y^2$? You know algebra? Charles.

CHARLES. One.

SERGE LANG. Yes. Why? Square x. What do you get when you square x? Figure it out. Dictate. What's the square of the numerator?

CHARLES. $1 - 2t^2 + t^4$.

SERGE LANG. That's right. And now suppose you add the y squared.

CHARLES. You get

$$\frac{1 - 2t^2 + t^4}{1 + 2t^2 + t^4} + \frac{4t^2}{1 + 2t^2 + t^4}.$$

SERGE LANG. That's it. That's good. That's exactly it. Both x and y are over the same denominator. So what happens if you combine them? Combine the numerators. What do you get? Charles.

CHARLES. $1 - 2t^2 + t^4 + 4t^2$, which is $1 + 2t^2 + t^4$.

SERGE LANG. That's right, Charles is right, because $-2t^2 + 4t^2$ is $+2t^2$. So you get

$$x^2 + y^2 = \frac{1 + 2t^2 + t^4}{1 + 2t^2 + t^4},$$

where the numerator is the same as the denominator. So what do you get finally, Charles?

CHARLES. You get 1.

SERGE LANG. That's right. You get $x^2 + y^2 = 1$. Now suppose you let t have special values. Try out some special values. Let's take some examples. Somebody give me a value for t.

A STUDENT. Two.

SERGE LANG. All right, take $t = 2$. What do you get?

$$x = \frac{1 - t^2}{1 + t^2} = \frac{1 - 2^2}{1 + 2^2} = \frac{-3}{5}.$$

A STUDENT. And $y = \dfrac{2t}{1+t^2} = \dfrac{4}{5}$.

SERGE LANG. Yes, so that's my a/c and b/c. I can take $a = 3$, $b = 4$, $c = 5$. We get the solution 3, 4, 5 that we had before. Give me another value for t.

A STUDENT. $t = 3$.

SERGE LANG. All right, that's as good as any. Now plug it in. What do you get?

[*Student dictates:*]

$$x = \frac{1 - 3^2}{1 + 3^2} = \frac{-8}{10}.$$

SERGE LANG. And what is y?

STUDENT. $y = 6/10$.

SERGE LANG. OK. That's not so good, that's not interesting because you are getting a multiple of 3, 4, 5. Let's do another example. Take another value of t.

A STUDENT. Seven.

SERGE LANG. All right, when $t = 7$, what do you get for x?

STUDENT. $1 - 49 = -48$, and $1+49 = 50$ so $x = 48/50$.

SERGE LANG. And what do you get for y?

STUDENT. $y = 14/50$.

SERGE LANG. Yes, so if you reduce it to lowest form, you get?

STUDENT. 24/25 and 7/25.

SERGE LANG. Yes, that's another you picked out earlier, 7, 24, 25. Try another example.

STUDENT. $t = 11$.

SERGE LANG. All right. Then you get?

STUDENT. $x = \dfrac{1 - 121}{1 + 121} = \dfrac{-120}{122}$ and $y = \dfrac{22}{122}$.

SERGE LANG. OK. And now if I reduce the fractions, what do you get?

STUDENT. $-60/61$ and $11/61$.

SERGE LANG. Ah, Ah! That's a new one! 11, 60, 61. Do you want to give me one more value?

STUDENT. 13.

SERGE LANG. OK. Let's do the computation. [*Serge Lang and students do the computation.*] If $t = 13$, we get

$$x = \frac{-168}{170} = \frac{-84}{85} \quad \text{and} \quad y = \frac{26}{170} = \frac{13}{85}.$$

So this is another one, 13, 84, 85. Now it's clear it will go on. Also you gave me values of t which are whole numbers, which are integers. You could have given me a fraction. Want to try a couple of fractions? [*Students nod.*] All right try a couple of fractions. Give me a value of t which is a fraction.

A STUDENT. 1/2.

SERGE LANG. Then x is what?

[*Student dictates:*]

$$x = \frac{1 - 1/4}{1 + 1/4} = \frac{3/4}{5/4} = \frac{3}{5} \quad \text{and} \quad y = \frac{2 \cdot 1/2}{5/4} = \frac{4}{5}.$$

Everybody sees this? [*Students agree.*] So I am getting back 3, 4, 5.

[*Serge Lang and the class go through a couple more examples with fractions, like* 1/7.] So it's clear I can substitute any fraction for t, and I get a, b, c; so I get a whole slew of solutions. Obviously I get infinitely many, just by substituting fractions for t, by substituting rational numbers for t.

Now we have made a lot of progress from a few minutes ago, because a few minutes ago, even using all your brainpower, you were still getting them one at a time and experimenting. And it was not clear whether you'd get infinitely many. Now we've made the progress that we can actually write down a whole slew of them. All right, let's call it a parametrization. Have you heard the word "parametrization"? t is called the parameter. We substitute values for t in perfectly well defined expressions. We've written down expressions, we substitute values for t, and we get solutions to the problem of solving $x^2 + y^2 = 1$ in rational numbers.

The next question is, have we got all the solutions or not? Do you think there are any other solutions besides those which you write down like this? This is quite a different type of question. It looks much more complicated. Do you see that we are able to write down a bunch of solutions. Are there any others? Have we missed any? Do you have any opinion?

A STUDENT. I don't know. Maybe we have missed some.

THE TEACHER. Is it possible that any of these could duplicate?

SERGE LANG. Yes, some duplicate. We have already seen that we get 3, 4, 5 from two of them. But now I'm dealing with the problem of all the solutions of $x^2 + y^2 = 1$. The problem of duplication can be discussed afterward. At the moment, I want to know: have we got them all? What do you think?

A STUDENT. I don't think so.

SERGE LANG. You don't think I've got them all?

STUDENT. You're using rational numbers?

SERGE LANG. Sure, I'm using rational numbers. Using the formulas

$$x = \frac{1-t^2}{1+t^2} \quad \text{and} \quad y = \frac{2t}{1+t^2}$$

and substituting rational numbers for t, do I get all the rational numbers (x,y) which satisfy the equation $x^2 + y^2 = 1$? How do you know that there isn't another formula which gives me numbers x and y which don't come from this particular way of setting them up?

STUDENT. I don't.

SERGE LANG. You don't. What do you think?

ANOTHER STUDENT. I don't know!

SERGE LANG. But you understand the question?

STUDENT. Yes.

SERGE LANG. [*Pointing to another student.*] What do you think?

STUDENT. I think there may be another formula.

SERGE LANG. You think there may be another formula? Well, I want to show you one solution that we didn't get,

$$x = -1, \quad y = 0.$$

This is certainly a solution. If $x = -1$, what is x^2?

A STUDENT. 1.

SERGE LANG. If $y = 0$, what is y^2?

STUDENT. 0.

SERGE LANG. Add 0 to 1 and you get 1. Now I claim I can never get $x = -1$ from my formula, or $y = 0$. Because if $y = 0$ what is t?

STUDENT. Zero.

SERGE LANG. Yes, $t = 0$. And if $t = 0$ what is x?

STUDENT. 1.

SERGE LANG. So x is not equal to -1. Therefore I cannot possibly get the solution $x = -1$ and $y = 0$. So I miss at least one solution. Have I missed any other?

STUDENT. When $x = 0$.

SERGE LANG. If $x = 0$, what is t?

STUDENT. Either 1 or -1.

SERGE LANG. Well, that's still OK. Suppose $t = 1$ or $t = -1$. I get a solution with $x = 0$. So that's now the question: are there any other solutions other than this one, $x = -1, y = 0$, and the ones given by the formulas? How many think it's a cute question? [*Laughter.*] How many don't give a damn? [*Strong laughter.*]

A STUDENT. Do you think x and y can both be zero?

SERGE LANG. No, if x and y are both zero, they won't add up to 1.

STUDENT. Oh, OK. Yeah.

SERGE LANG. $x^2 + y^2$ has to be 1. I repeat the question, have I got all the solutions? [*The students give signs of thinking.*]

All right, now I'll state the theorem.

Theorem. All solutions of $x^2 + y^2 = 1$ in rational numbers x, y are obtained from the formulas

$$x = \frac{1-t^2}{1+t^2}, \quad y = \frac{2t}{1+t^2}$$

by giving special rational values of t, except the solution $x = -1, y = 0$.

That is the only one I have missed. All the others have been obtained. Do you know how to prove it? If I can prove it, then I have completely solved the problem of solving $x^2+y^2 = 1$ when x and y are rational numbers. This is a complete solution of the problem I posed at the beginning.

I am now going to give the proof. Suppose I have a solution. I want to show that there exists a value of t, which is a rational number, such that

$$x = \frac{1-t^2}{1+t^2} \quad \text{and} \quad y = \frac{2t}{1+t^2}.$$

If you try to discover how to get the t, you will find yourself the patterns that–what's your name?

STUDENT. Selim.

SERGE LANG. The patterns that Selim gave at the beginning. But of course I gave you solutions so that I can give the complete proof before the end of the hour. We didn't have time to experiment. You would have to experiment, maybe for two days, before you would find how to do it. One would have to systematize the patterns Selim was trying to find at the beginning, experiment, and maybe after some time you would get the proof. Now I'll show you the proof by writing it down. It's a "let". You know what a "let" means? [*Laughter.*] "Let" means you had a brainstorm and you write down the solution, and you prove afterward that it's the solution. So it's a "let". I let

$$t = \frac{y}{x+1}.$$

It's a "let". Now either this will work or it won't. So what I have to show, is that if I let t equal y over $x+1$, as in the picture:

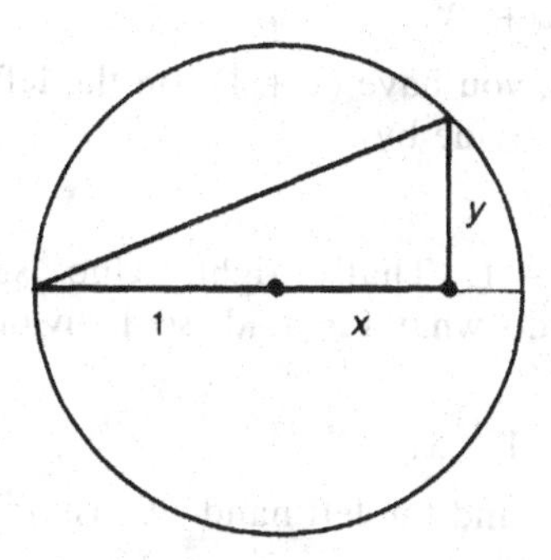

then x and y are given by the formulas. From here on, it's just a question of algebraic manipulation. If I let $t = y/(x+1)$ and x, y are rational numbers, then t is a rational number, right? If you add fractions, you get a fraction. If you multiply fractions, you get a fraction. If you take the quotient of two rational numbers, you get a rational number. So t is a rational number. Now I want to show that

$$x = \frac{1-t^2}{1+t^2} \quad \text{and} \quad y = \frac{2t}{1+t^2}.$$

Observe that in the formula $t = y/(x+1)$, the denominator $x+1$ does not make sense if $x = -1$. So I have to omit this case. But otherwise, it does make sense. Now it's just a question–you know–like in the kitchen. You cook a little, and it comes out. We have

$$(x+1)t = y.$$

If you square this you get

$$(x+1)^2t^2 = y^2.$$

And y^2 is what in terms of x? If $x^2 + y^2 = 1$, then y^2 equals what?

SELIM. $1 - x^2$.

SERGE LANG. Yes, and $1 - x^2$ can be factored. How?

SELIM. $1 + x$ times $1 - x$.

SERGE LANG. Yes, you are doing very well. You guys are very good, very much at ease with algebra. So I have

$$(x + 1)^2t^2 = (1 + x)(1 - x).$$

Now you should have an irresistible impulse to do something, which is what? [*Pointing to a student.*] What is your name?

STUDENT. Uno.

SERGE LANG. All right, Uno, what do you want to do now?

UNO. Divide by $(x + 1)^2$.

SERGE LANG. No, you have $(x + 1)^2$ on the left but only $x + 1$ on the right. So you have to divide by . . . ?

UNO. $x + 1$.

SERGE LANG. $x + 1$. That's right. Uno says divide by $x + 1$. [*Laughter.*] I always do what I'm told, so I divide by $x + 1$. What do I get?

UNO. $(x + 1)t^2 = 1 - x$.

SERGE LANG. Yes, and the left hand side is $xt^2 + t^2$, so I have

$$xt^2 + t^2 = 1 - x.$$

Now you solve for x. What do I get?

$$x(1 + t^2) = 1 - t^2.$$

And then I can get

$$x = \frac{1 - t^2}{1 + t^2}.$$

Then

$$y = (x + 1)t,$$

and you can use algebra again to see that $y = 2t/(1 + t^2)$. This is just what we wanted to prove.

STUDENT. How do you find this proof?

SERGE LANG. A good mathematician, fooling around for two days, would be able to get the "let", write down the "let", and then do some algebraic twiddling to prove that the "let" works. That's the difference between a good mathematician, and someone who would not find the "let", and would not find the proof. Some of you can, and some of you cannot, find the solution. That's the difference between doing research mathematics and not doing it. But that's what research mathematicians do. They start the way Selim did. They try to experiment, and after a while, one sees what works, and one finds the proof that it is a solution. That's the answer to your question.

Comments

The period ended before there was time to expand on this. Pressed on several occasions concerning the history of the formulas and the "let" which I introduced somewhat abruptly, I finally inquired into the history of the formulas. G. Lachaud informed me that it took centuries before the formulas took the final form which I have given here, with contributions by several mathematicians. Euclid, or mathematicians of his time, knew three centuries BC that you could find Pythagorean triples a, b, c by using the formulas

$$a = m^2 - n^2, \qquad b = 2mn, \qquad c = m^2 + n^2,$$

where m, n are integers. Check using algebra that you indeed get $a^2 + b^2 = c^2$ from these formulas. Then you can give m, n arbitrary values in integers to get solutions. Mathematicians at the time of Euclid did not like fractions, and preferred to deal entirely with integers. Only much later, Diophantus (three centuries AD) worked with fractions, and knew that if you divide the above formulas by $m^2 + n^2$, and then put $t = n/m$, then you find the formulas

$$\frac{a}{c} = \frac{1-t^2}{1+t^2} \quad \text{and} \quad \frac{b}{c} = \frac{2t}{1+t^2}.$$

So it took six centuries between Euclid and Diophantus (who worked at Alexandria in Egypt, but was a Greek mathematician) to find the formulas as I wrote them down at the beginning, to get infinitely many solutions of the equation $x^2 + y^2 = 1$ in rational numbers.

Diophantus himself did not raise the question whether these formulas give all solutions in rational numbers. For this one had to wait another six centuries. In the 10th and 11th century, Arab mathematicians like Al Khazin raised this further question, and solved it by algebra, essentially as we did it today. Since the algebra is about at the same level as that known to Diophantus, we now see after the fact that if Diophantus had raised the question, he would have been able to solve it.

This history shows that a complete solution resulted from the work of mathematicians ranging over thirteen centuries. Since then, mathematicians (including me) have copied the solution, as I did today.

It may interest you to pursue the idea of Selim, who noticed the pattern that the odd integers 3, 5, 7, 9, 11 seem to occur systematically at the beginning of Pythagorean triples of integers, which are solutions of $a^2 + b^2 = c^2$. Is it true that every odd integer will be the first integer in such a triple? If it is true, prove it. If it is false, give an example of an odd integer which does not occur.

There was also no time to deal with the question raised by one of the teachers (and sometimes raised by students) whether and how there are duplications in the solutions coming from the formulas. This is in fact quite easy to do.

For example, suppose t ranges from 0 to 1, so t increases from 0 to 1. Then $1 - t^2$ decreases from 1 to 0; and $1 + t^2$ increases from 1 to 2. Hence the quotient

$$x = \frac{1 - t^2}{1 + t^2}$$

decreases from 1 to 0, and there are no duplications of x for this range of values of t.

What happens when t ranges over all numbers greater than 1? Write $t = 1/s$ and figure it out for yourself. Also figure out what happens when t is a negative number. And finally, figure out what happens when you write $t = m/n$, whether in lowest form or not to answer completely the question about duplication of solutions, or the possibility of getting integer multiples of solutions a, b, c for $a^2 + b^2 = c^2$.

Today, students at the age of 16 know something of coordinates and the representation of the circle by the equation $x^2 + y^2 = 1$. Neither Euclid, Diophantus, nor Al Khazin knew about that. The idea of using coordinates arose only in the 16th and 17th century. You will notice that although I drew a circle and interpreted the problem of finding solutions of the equation in terms of finding "rational points on a circle", the figure was not used in the arguments. I argued completely algebraically, much as the ancients would have done. However, today it is also useful to give a geometric interpretation of the algebraic steps. In particular, when I let

$$t = \frac{y}{x+1},$$

what is the geometric interpretation of t? If I had another hour, I would lead students to realize that the equation

$$y = t(x+1)$$

is the equation of a straight line, with slope t.

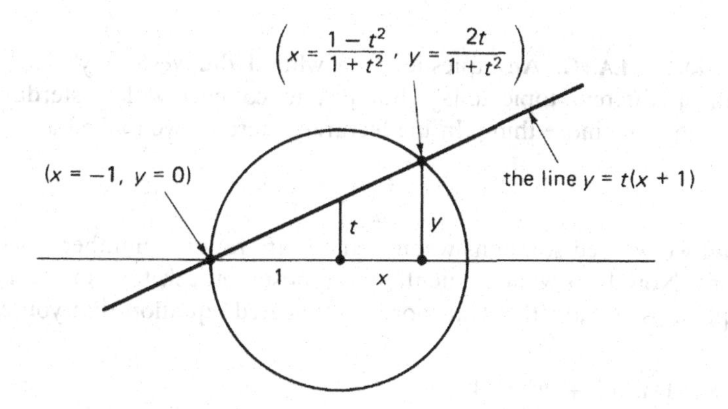

This line passes through the point $x = -1, y = 0$. The point where the line intersects the circle is precisely the point given by

$$x = \frac{1-t^2}{1+t^2}, \quad y = \frac{2t}{1+t^2}.$$

This is the geometric interpretation of the part of the proof showing how to get back these formulas for x and y in terms of $t = y/(x+1)$.

Infinities

The following talk is the second given to an 11th grade class in a high school in the suburbs of Toronto in Spring 1982. The talk lasted about 50 minutes. Although the first few minutes are spent discussing a topic extending that of the first talk, I have preserved the timing and balance of the talk intact.

SERGE LANG. Any questions on what I did yesterday? No? I'll deal with a different topic today, but just to connect with yesterday, let me mention one more thing. In the lecture yesterday, we looked at

$$x^2 + y^2 = 1.$$

And we wanted solutions when x and y are rational numbers, that is fractions. Now I do want to mention–suppose you ask for more complicated equations. What's the next more complicated equation that you can think of?

SELIM. $x^3 + y^3 = 1$.

SERGE LANG. That's right, you can either have this, or something like

$$x^3 + y^2 = 1.$$

Either one or the other. Now do you think there are infinitely many solutions of this equation in rational numbers?

SELIM. I don't know.

SERGE LANG. You don't know?

SELIM. There should be.

SERGE LANG. Oh you think so?

SELIM. Yeah. There were infinitely many yesterday.

SERGE LANG. There were infinitely many for $x^2 + y^2 = 1$. What about for $x^3 + y^2 = 1$?

SELIM. I can see at least four. Right away.

SERGE LANG. Are you sure? Which four?

SELIM. $x = -1, y = 1$.

SERGE LANG. [*Interrupting.*] What?

SELIM. Eh, no [*he computes in his head and partly out loud*].

SERGE LANG. It's not so clear, is it?

SELIM. I'm not sure exactly.

SERGE LANG. Ah, now you are not sure. That's right. Well, the situation is infinitely more difficult for equations with a cube than only with squares. And I think the only solutions are essentially $x = 0, y = 1$, or $x = 1, y = 0$, or $x = -2, y = 3$. There is no other, unless you allow $y = -1$ or $y = -3$, because y^2 gets rid of the minus sign. That's quite hard to prove. Quite hard. And similarly for higher powers. It is exceedingly difficult to handle such equations the moment the degree is bigger than 2. It's a very famous unsolved problem, that nobody knows how to solve. If you take an arbitrary n, and the equation

$$x^n + y^n = 1,$$

show that only solutions are with $x = 0, y = 1$ or $x = 1, y = 0$, except for putting a possible minus sign. That's a very difficult problem, which is called the Fermat problem, and nobody knows how to do it.

One knows how to do it for low values of n, like $n = 3, 4, 5, 6, 7$. For a while, for low values of n, people can hack it out, by brute force, using ad hoc methods. But to give a proof in general, nobody knows how to do it. It's what's called an unsolved problem in mathematics. That's what mathematicians do, by the way. They pick nice unsolved problems, that they get a high on, then they work on them. It gives them a high to work on unsolved problems. Do you think it would give you a high to work on an unsolved problem of mathematics?

A STUDENT. Sure. [*Laughter.*]

ANOTHER STUDENT. We'll have it ready for you next time.

SERGE LANG. You'll have it ready for me next time? If you do, you'll make it into the history books. [*Laughter.*] Want to make it into the history books?

Actually, I was just mentioning this, now let me get on to the topic I want to discuss today. I want to discuss questions of infinity. For example, take the integers,

$$1, 2, 3, 4, 5 \text{ and so on}.$$

They go on indefinitely. And there are infinitely many of them. Now suppose I take the odd numbers. You know what the odd numbers are?

A STUDENT. Those that are not even. [*Laughter.*]

SERGE LANG. All right, what are the even numbers?

THE STUDENT. 2, 4, 6, 8, ...

SERGE LANG. All right, I write them down,

positive integers: 1 2 3 4 5 6 7 . . .
even numbers: 2 4 6 8 10 12 14 . . .

Observe that you can make them correspond with the positive integers. So there are no even numbers missing in the correspondence. I can line up the even numbers, a first one, second one, third one, and so on, so I won't miss any of them. When that happens, I say that the even numbers are denumerable. "Denumerable" means that I can enumerate them. A first one, second one, third one, and so on, to take them all into account. What about the odd numbers? Are they denumerable? [*Serge Lang points to a student.*] What's your name?

STUDENT. Ted.

SERGE LANG. OK, Ted, are the odd numbers denumerable?

TED. I'm not sure what you mean.

SERGE LANG. "Denumerable" means that I can list them, a first, second, third, etc. so that I don't leave out any of them. Here are the odd numbers:

odd numbers: 1 3 5 7 9 11 13 15 . . .

For the even numbers, 2 is the first, 4 is the second, 6 is the third, 8 is the fourth, 10 is the fifth, and so on. Which is the tenth even number?

TED. 20.

SERGE LANG. That's right. Which is the thirtieth?

TED. 60.

SERGE LANG. Yeah. Have I missed any?

TED. No.

SERGE LANG. So I have enumerated the even numbers. Which is the first odd number?

TED. One.

SERGE LANG. Which is the second odd number?

TED. 3.

SERGE LANG. Which is the third odd number?

TED. 5.

SERGE LANG. And then?

TED. The fourth is 7, then the fifth is 9, and so on.

SERGE LANG. Right. So I have a one to one correspondence between the odd numbers and the positive integers. They correspond, it's one-one.

One integer corresponds to one odd number. 1 corresponds to 1, 2 corresponds to 3, 3 corresponds to 5, 4 corresponds to 7, 5 corresponds to 9, and so forth. I have enumerated the odd numbers.

$$\begin{array}{ccccccccc} 1 & 2 & 3 & 4 & 5 & 6 & 7 & 8 & \dots \\ 1 & 3 & 5 & 7 & 9 & 11 & 13 & 15 & \dots \end{array}$$

Have I missed any odd numbers in this way?

TED. No.

SERGE LANG. So then I would say that the odd numbers are denumerable. By the way, could you describe this correspondence by a formula? If I call n the n-th integer, like the first one, second one, third one, n-th one, what will be the n-th even number?

TED. $2n$.

SERGE LANG. That's right, the n-th even number is $2n$. The n-th odd number would be what?

TED. $2n - 1$.

SERGE LANG. Yes. That's right, good for you. You said it, the n-th odd number is $2n - 1$. So that's how I enumerate the odd numbers and the even numbers. Even though the odd numbers and the even numbers form part of all the positive integers, I call still make this one-one correspondence between them, and enumerate them.

Now I have a question. What's the next kind of number you know of? The rational numbers, yes? That's ordinary fractions.

Question. Are the rational numbers denumerable?

Do you understand what I mean? Take the rationals, say the positive ones. Forget about the negative ones. So it's numbers like 2/3, 7/5, 13/37, fractions. Can I enumerate them? Any integer is of course a fraction. I can write $1 = 1/1$, $2 = 2/1$, $3 = 3/1$. So certainly the positive integers constitute part of all the positive rational numbers. Now can I take all the rational numbers, and line them up, so there is a first one, a second, a third, and so on, and so that I don't miss any of them?

How many say yes? [*A few hands go up.*]

How many say no? [*Some hands go up. Mostly nobody says anything.*]

How many keep a prudent silence? [*Laughter.*]

All right, let's see those who say no. Who said no? Did you say no?

STUDENT. I said neither.

SERGE LANG. You said neither? [*Several students talk simultaneously.*] All right, you [*Serge Lang points to a student*]. [*Laughter.*] What did you say?

STUDENT. No I don't think so, but I don't know.

SERGE LANG. Why? Why not? Can you give a reason for it?

STUDENT. No.

ANOTHER STUDENT. Well, if anybody says you can enumerate them, you give me two, and you can find another one between them.

SERGE LANG. Ah! that is in the usual ordering. Sure, if you take all the rational numbers on the number line, and if you take two rational numbers, you can find another one in between.

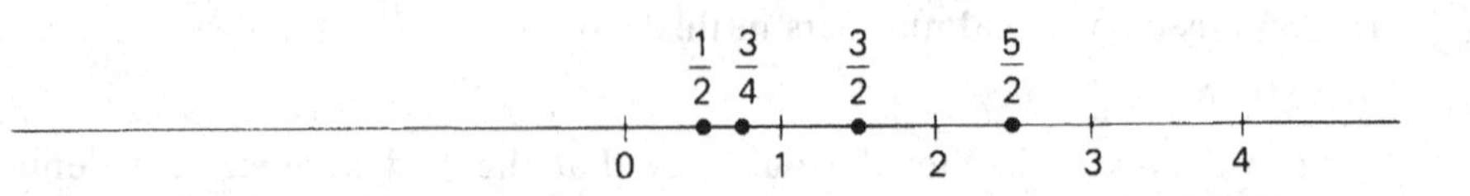

But I didn't say to take the usual ordering. Suppose you change the ordering?

STUDENT. I think in any ordering you could do the same.

SERGE LANG. Any ordering?

STUDENTS. The fractions are infinite.

SERGE LANG. Oh, but the even numbers were infinite too!

STUDENT. No, but what I am saying, is the fractions between two integers are infinite.

SERGE LANG. Do you think that's an argument? It's true, if I take two fractions in the usual ordering, there will be another fraction in between. Now can I line up the fractions in a different ordering, so there is a first, a second, a third, and I don't miss any?

ANOTHER. You have an infinite number of denominators, right?

SERGE LANG. Yes I do.

STUDENT. And each denominator has infinitely many numerators. So how can you line it up when you don't know how high this is going to go?

SERGE LANG. My answer to you is: you ask me how can I line them up. Would my failure to line them up be due to my stupidity . . .

STUDENT. No.

SERGE LANG. . . . or because of the mathematics?

STUDENT. Because there is an infinite number of denominators. Like the numerator of the fractions can keep changing and changing if I made the denominators bigger.

SERGE LANG. That's true. But that's still not an argument. How do you know . . .

STUDENT. How can you line them up if you can keep changing the denominator? Like for those . . .

SERGE LANG. Well, suppose I was clever enough! You haven't given an argument. I mean, I don't know at the moment. I haven't told you what's true or not. I'm simply questioning you. How do you know if it's built in the structure of the mathematics, or if it's built in to the accidental thing that I am not clever enough to do it?

[*Long silence.*]

You have to give a proof. If you assert you cannot do it, you will have to prove it. [*A hand goes up.*] Yes?

STUDENT. [*The tape cannot be heard clearly. The student appears to make a reference to the step by step ordering of the even and the odd numbers. One can hear faintly: the smallest step could be one.*]

SERGE LANG. It's true that the even and odd numbers are arranged by steps. Maybe in dealing with the rational numbers I cannot deal with something that you call the steps. I have to deal with something else. All you have said is that the method which I used for the even and odd numbers does not apply to the rational numbers. That's all you have said so far. How do you know that if I'm clever enough, I can't find another method? [*A hand goes up.*] Yes?

STUDENT. Well you take 1, 2, 3, 4, 5, 6 and so on. You put them all, the whole thing, you put them over 1. And then you do it—you put them over 2, and you do it over 3. Then the denominators, they go up . . .

SERGE LANG. Yes, they go up. That's very good. What's your name?

STUDENT. Gary.

SERGE LANG. So what Gary is proposing is this. Let's line up all the fractions according to their denominators. By the way, what is your name? [*Pointing to the student who had raised the question about denominators.*] What's your name?

STUDENT. Ken.

SERGE LANG. Ken also said what happens when the denominators grow bigger. So both Ken and Gary said something similar. We have the fractions, say with denominator 1:

$$\frac{1}{1}\ \frac{2}{1}\ \frac{3}{1}\ \frac{4}{1}\ \frac{5}{1}\ \frac{6}{1}\ \frac{7}{1}\ \frac{8}{1}\ \cdots \text{ and so on.}$$

Now they were mentioning the denominators which grow bigger.

GARY. With 2.

SERGE LANG. Yes, so you have

$$\frac{1}{2}\ \frac{2}{2}\ \frac{3}{2}\ \frac{4}{2}\ \frac{5}{2}\ \frac{6}{2}\ \frac{7}{2}\ \frac{8}{2}\ \cdots \text{ and so on.}$$

Then you have

$$\frac{1}{3}\ \frac{2}{3}\ \frac{3}{3}\ \frac{4}{3}\ \frac{5}{3}\ \frac{6}{3}\ \frac{7}{3}\ \frac{8}{3}\ \cdots \text{ and so on.}$$

Then, what would be the next one? Gary.

GARY. $\frac{1}{4}\ \frac{2}{4}\ \frac{3}{4}\ \frac{4}{4}\ \frac{5}{4}\ \frac{6}{4}\ \frac{7}{4}\ \frac{8}{4}\ \cdots$ and so on.

SERGE LANG. So I have lined up the fractions, horizontally for those with the same denominator, and vertically with the denominator getting bigger.

$$\frac{1}{1}\ \frac{2}{1}\ \frac{3}{1}\ \frac{4}{1}\ \frac{5}{1}\ \frac{6}{1}\ \frac{7}{1}\ \frac{8}{1}\ \cdots$$

$$\frac{1}{2}\ \frac{2}{2}\ \frac{3}{2}\ \frac{4}{2}\ \frac{5}{2}\ \frac{6}{2}\ \frac{7}{2}\ \frac{8}{2}\ \cdots$$

$$\frac{1}{3}\ \frac{2}{3}\ \frac{3}{3}\ \frac{4}{3}\ \frac{5}{3}\ \frac{6}{3}\ \frac{7}{3}\ \frac{8}{3}\ \cdots$$

$$\frac{1}{4}\ \frac{2}{4}\ \frac{3}{4}\ \frac{4}{4}\ \frac{5}{4}\ \frac{6}{4}\ \frac{7}{4}\ \frac{8}{4}\ \cdots$$

Now can I enumerate them? I have lined them up the way you wanted me to line them up. Now can I make a list? A first, second, third, fourth, so that I don't miss any of them?

[*Silence. A hand goes up.*] Yes?

A STUDENT. No, because the denominators and numerators are getting bigger.

SERGE LANG. Is that a reason? You can make the list any way you want.

THE STUDENT. You have to use [*tape is not clear*].

SERGE LANG. I can use anything you want. You can make the list any way you want. As long as there is a first, a second, a third and so on. I can use any method you want, as long as they are lined up and I get all of them. All you have to do is to be clever. Do you think it's a nice question? [*Smiles, some students say yes.*] Do you think it's a dumb question? [*Silence.*] Well, you are paying attention.

A STUDENT. Well you take 1/4; and then you take diagonally across, it would be 1/3 and 2/4; then you have 2/3 . . .

SERGE LANG. Ah, Ah! You got it! Heh!? He got it! What's your name?

STUDENT. Sunil.

SERGE LANG. OK, I'll do exactly what Sunil says. Let's start with this corner [*points to the upper left hand corner, to the fraction* 1/1]. This is my first, 1/1. This is the second, 2/1. This is the third, 1/2. This is the fourth, 1/3. This is the fifth, 2/2. Sixth, 3/1. Seventh, eighth, ninth, tenth, . . .

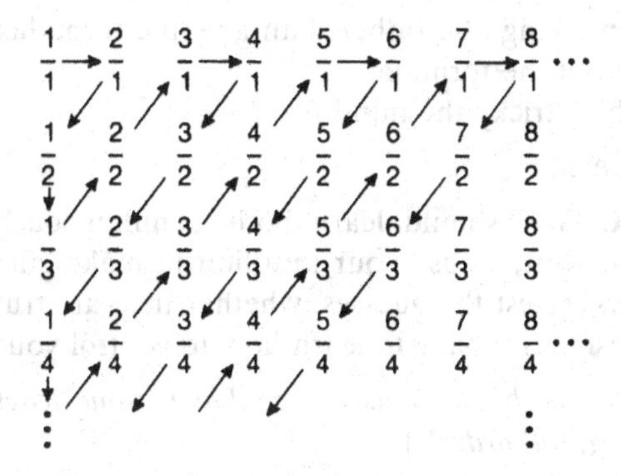

eleven, twelve, thirteen, fourteen, fifteen, sixteen, seventeen, eighteen, . . .
Now do you see it? Have I enumerated them? Am I going to miss any of them by this method?

STUDENT. No.

SERGE LANG. Eventually, I will encounter every one of them. I will meet all of them, and there will be a first, a second, a third, a fourth, and so on. So in fact they are denumerable. You understand the proof? That's the theorem. The answer is yes.

Theorem. The positive rational numbers are denumerable.

And the proof is right there. I have given you the enumeration. You agree that I have enumerated them?

STUDENT. Yeah.

SERGE LANG. So what about the people who said I couldn't do it? What happened, heh? I mean, it's tricky, the mind. It's tricky.

A STUDENT. Is there an equation for that? Can you give an equation?

SERGE LANG. Yes, you can, but it's more complicated. I would have to give the exact recipe how to do it, by an equation. In fact, I don't know offhand, how to write the equation. Why don't you do that as homework? [*Strong laughter.*] Try to think through . . . [*Very strong laughter.*] No, it's not so bad, it's not so bad. Well, it will take you . . . depending how fast your mind works, it might take you from five minutes to an hour. To give the equation, to give the relationship, as we did for the even and odd numbers: $2n$ and $2n-1$. These were the formulas. Now how do you give the formulas in this case, for the n-th rational number, it's an exercise.[1] But it's a different question, and I want to do something else, and if I start working on the formulas for this, I won't have time to do something else.

[1] Thinking about this later, I don't see how to give a single formula. I would do it only with several formulas, to fit the exact enumeration as on the diagram.

I'd rather do something else rather than give the formulas for this one. So I'll let you work out the formulas.

But you see how tricky the mind is?

STUDENT. Yeah.

SERGE LANG. You should learn both, simultaneously: to trust your intuition, to distrust it; to use your intuition to make guesses, to use your analytical powers to test the guesses, whether they are true or false. All of this simultaneously. You have to learn how to control your brain.

[*Simultaneous talk by students, Serge Lang, some teachers in the back who say "it's a big, tall order".*]

You think that's a big, tall order? It's not that bad.

So now we go to the next question. We decided that the rational numbers are denumerable. Although it looks as if there are a lot more rational numbers than integers, we have just enumerated them. Now do you know what real numbers are? The number line. You know the number line? All the numbers on the line.

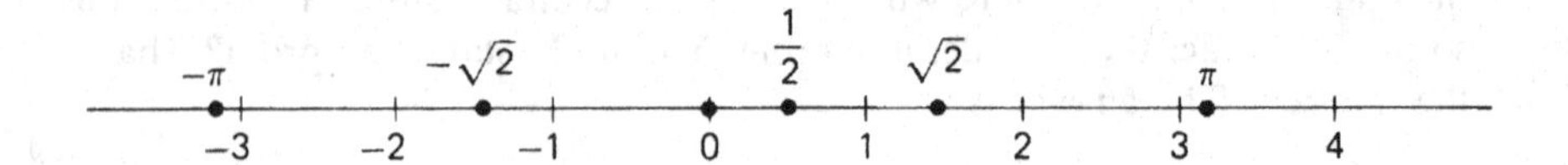

STUDENTS. Yes we know. [*Several talk simultaneously.*]

SERGE LANG. Do you think those numbers are denumerable? Can you line them up so there is a first, a second, a third, a fourth, and so on? How many of you say yes? That's my question. Let me write down the question.

Are all the numbers on the number line denumerable?

By the way, we might as well limit ourselves to the numbers between 0 and 1. For simplicity. So how many say yes? Raise your hands. How many say no? [*Laughter.*] How many keep a prudent silence? [*Almost nobody raised their hands either time. Laughter when no hands are raised.*]

SERGE LANG. Oh, you guys are getting careful! What do you think? Not so clear, heh. After you got caught a few minutes ago, with the rationals, one has to be careful. What do you think? [*Serge Lang points to a student.*]

A STUDENT. I think you can show they are denumerable.

SERGE LANG. They are denumerable? Why?

STUDENT. You could use the same reason as the other one.

ANOTHER STUDENT. No, I don't think so.

SERGE LANG. Well, let's see. You look at all the numbers between 0 and 1.

0 1

KEN. Between 0 and 1 you have infinitely many of them.

SERGE LANG. But there is also an infinite number of fractions, between 0 and 1, and you can enumerate the fractions. [*A hand is raised.*] Yes?

STUDENT. I wonder if you could use denominators again.

SERGE LANG. No there are no denominators. All the numbers x between 0 and 1 . . . What is by the way a "number" on the number line? What's the easiest way of representing a number?

KEN. Decimals.

SERGE LANG. Decimals, OK. So I take all the decimals. For example.

$$0.13251986127824 \ldots \text{ and so on.}$$

That's a typical infinite decimal. So that's what my numbers are. Decimals like that. I take the decimals as the definition of numbers. A number is an infinite decimal. Now, can I line up the decimals so there is a first decimal, a second decimal, a third decimal, and so on, so that I don't miss any? Can I do that? That's the question.

A STUDENT. Each number is an infinite decimal.

SERGE LANG. Yes, that's right. Each number has an infinite decimal. That's why I put the dot, dot, dot . . .

THE STUDENT. What about π? π does not have a decimal value.

SERGE LANG. Sure, $\pi = 3.14159\ldots$ You know there is an infinite decimal which gives the value for π. The value is itself the infinite decimal. That's the value of the number. You know π does not stop after 3.14. Don't you know that? I mean, there is a 159 [*Laughter*]. It goes on like this:

$$3.141592653589\ldots$$

[*Serge Lang writes many of the decimals. Laughter.*]

TEACHER. You're using a . . .

SERGE LANG. Shhhh!

TEACHER. OK, I won't say.

SERGE LANG. Don't give away my secret![2]

---

[2] The secret is a poem in French: Que j'aime à faire apprendre un nombre utile aux sages! Immortel Archimède, artiste, ingénieur, qui de ton jugement peut priser la valeur? Pour moi ton problème eut de pareils avantages.

TEACHER. He is working it out in his head, that's right.

[*Laughter. Talk by students. He's doing it in his head! How are you doing it? What are you thinking?*]

SERGE LANG. So a number is an infinite decimal. Can I line up the numbers so there is a first one, second one, third one, and so forth? What do you say?

STUDENT. No.

SERGE LANG. He says no.

STUDENT. With the rationals, you had a place to start. [*Tape cannot be made out.*] Now you don't have a place to start.

SERGE LANG. Yes, but again, all that you are saying is that the method which applied to the rationals will not apply to the real numbers. The people who started to give reasons why I could not do it for the rationals: all they were saying is that the method which applied to the even numbers did not apply to the rationals.

STUDENT. Probably there is.

SERGE LANG. Oh yeah, probably there is?

STUDENT. So far you used different methods; you first used one method, then you changed the method; probably a different method would do it for all the numbers.

SERGE LANG. That's a very weak argument. [*Laughter.*] Because the argument is based on psychology, and I am asking you to deal with mathematical problems. Not psychological ones. [*Laughter.*] So if you start basing your mathematical intuition on my psychology, [*Laughter.*] you're going to have a hard time with it. That's dangerous. Think again.

[*Students talk among each other.*]

SERGE LANG. All right, I'll tell you the theorem.

Theorem. The infinite decimals are NOT denumerable.

That's the theorem. You cannot do it. And I'll prove it. Well, suppose you could enumerate the infinite decimals. Suppose there is a first one. Suppose you make a list of them. So there is a first one, a second, third, fourth . . . I have to write them as infinite decimals. I can't write an infinite decimal completely. I have to use some means of expression for the decimals. Suppose we have the first decimal

$$\text{first:} \quad .a_{11}a_{12}a_{13}a_{14}a_{15}a_{16}\ldots$$

So each a represents an integer between 0 and 9, each a_{11}, a_{12}, a_{13}, and so on, is an integer between 0 and 9. Suppose there is a second one, which I write second: $.a_{21}a_{22}a_{23}a_{24}a_{25}a_{26}\ldots$ and so on.

third: $.a_{31}a_{32}a_{33}a_{34}a_{35}a_{36}$,,, and so on.

Now the fourth one, how would I denote it?

GARY. $.a_{41}a_{42}a_{43}a_{44}a_{45}a_{46}$ and so on.

SERGE LANG. Yes. And what would be the n-th one? How would I denote the decimal for the n-th one? Gary.

GARY. $.a_{n1}a_{n2}a_{n3}a_{n4}\cdots$

SERGE LANG. So suppose I have such an enumeration of my infinite decimals:

first one: $.a_{11}a_{12}a_{13}a_{14}a_{15}\ldots$

second one: $.a_{21}a_{22}a_{23}a_{24}a_{25}\ldots$

third one: $.a_{31}a_{32}a_{33}a_{34}a_{35}\ldots$

....

n-th one: $.a_{n1}a_{n2}a_{n3}a_{n4}a_{n5}\cdots$

....

I have to show that there is one number, one infinite decimal, that is not in the list. If I show that, I have proved the theorem. Because I have shown that if you make any list, I can always find one number that is not in the list. So I have to give you a method for finding one which is not in the list. How do I find one which is not in this list?

[*Silence.*]

I have to give you a recipe, how to find one which is not in the list. Any one which is not in the list is going to be different from those in the list. I do it this way. I pick a number b_1, an integer between 0 and 8 such that b_1 is not equal to a_{11}. I mean, a_{11} is whatever it is. Suppose a_{11} is 2. Pick the first number b_1 to be 7. Whatever you want, except 2. Now pick the second decimal. Pick a number b_2, any integer between 0 and 8, such that b_2 is not equal to a_{22}. Next pick an integer b_3 such that b_3 is not equal to a_{33}. What's my next step?

STUDENT. Pick an integer b_4 such that b_4 is not equal to a_{44}.

SERGE LANG. That's right. And in general, pick b_n such that b_n is not equal to–what?

STUDENT. a_{nn}.

SERGE LANG. Perfect. And now I form the following decimal:

$$.b_1b_2b_3b_4\ldots b_n\ldots \text{and so forth}.$$

Can this decimal be equal to any of the ones which I have listed here?

STUDENTS. No.

SERGE LANG. Why not?

STUDENT. Because the numbers are different. [*The tape is not clear, but the student has the right idea.*]

SERGE LANG. That's right. Let's say it a little more precisely. Suppose this decimal

$$.b_1 b_2 b_3 b_4 \dots b_n \dots$$

is equal to one of those which I listed. Then it would be in this list, somewhere. Which one? Say the n-th one. But it can't be the n-th one, because the n-th one is

$$.a_{n1} a_{n2} a_{n3} \dots a_{nn} \dots$$

and it is different from the n-th one because we have made it different, by picking b_n different from a_{nn}. So the decimal $.b_1 b_2 b_3 b_4 \dots$ cannot be equal to any decimal in the list.

So what I have shown is that if you have a list to begin with, I can always find a decimal which is NOT in the list. Therefore there could not be a list to begin with, of all the decimals. Do you understand the argument? Selim, what do you say?

SELIM. Yes.

A TEACHER. But this rests on the assumption that you are able to find a b_1 which is not equal to a_{11}.

SERGE LANG. a_{11} is only one number. Suppose a_{11} is 2. Let b_1 be equal to 3. Suppose a_{22} is 5. Let b_2 be equal to 7. Suppose a_{33} is 1. Let b_3 be equal to 8. Suppose a_{44} is 4. Let b_4 be equal to 5. All I have to do is to make the b's different one at a time.

A STUDENT. Why did you take the b's between 0 and 8?

SERGE LANG. I took them between 0 and 8 for a technical reason. If you look at a number like this:

$$0.1399999999999 \dots$$

which goes on like this, with all 9's at the end; and you take

$$0.1400000000000 \dots$$

like that; then in fact, these two numbers are equal, and I want to avoid this kind of ambiguity which occurs in decimal expansions. So it's just a very minor technical point. If I hadn't mentioned it, you would not even have noticed it. I mentioned it for my own peace of mind. But if I had not mentioned it, I don't think you would have picked it up. That's the only reason. It's the only kind of ambiguity that can occur. If the same number has two different decimal expansions, the only difference can be that you have all 9's at the end of one of them, and all 0's at the end of the other. Otherwise, that's the only way you can have two different expansions for the same number. And that's the only reason I picked them between 0 and 8. Otherwise, forget it.

So you agree that the set of all infinite decimals is not denumerable? So it was pretty, yeah? The rational numbers look as if there are a lot more than the integers. Your first thinking was that–for most of you–the rational numbers are not denumerable. But they are, even though the denominators get bigger. Then I caught you psychologically with the real numbers–caught some of you–thinking, well, since it's going like that, they will also be denumerable. And then finally I pulled the rug from under you [*Laughter*] and showed that the real numbers are not denumerable.

So, this is what I wanted you to see. And you see, we are using very little here, about mathematics. We are just using the notion of correspondence, enumeration. We are not using arithmetic, addition, or multiplication, or anything like that. It's all just making things correspond to each other.

Postscript

What follows are lengthy excerpts from a discussion between Patricia Chwat, a teacher in a Paris High School; Jean Brette, director of the Mathematics Section of the Palais de la Découverte (the main science museum in Paris); Stephane Brette (14 years old), his son; and of course, Serge Lang.

The discussion brings up various points concerning the teaching of mathematics, and was transcribed from a videotape by Patrick Huet, of the Paris Research Institute for the Teaching of Mathematics (in French: IREM, Institut de Recherche pour l'Enseignement des Mathématiques). As far as I can tell, the points raised here are valid for all the countries that I know of, and provide an appropriate conclusion for this book.

Serge Lang

JEAN BRETTE. We just saw a math course you gave to 9th grade students in a Paris high school. Maybe we can recall first how it all started.

SERGE LANG. Where, at the Palais?[1]

JEAN BRETTE. No, in class. Well, of course, it all started at the Palais de la Découverte. Two years ago you gave us a talk there, and I asked a friend of mine if we could go and do some mathematics with the kids in his high school class. Last year, you came back for another talk, again at the Palais, and there was a polemic about teaching in high schools, in France and elsewhere, what you thought about the whole situation, and you recalled your experience about some talks you gave in a school in Canada.

SERGE LANG. Yes, that's right.

JEAN BRETTE. And after this discussion, Miss Chwat asked you to . . .

[1] Serge Lang gave three talks at the Palais de la Découverte, the main science museum in Paris, for three consecutive years from 1981 to 1983. These talks were first published in the Journal of the Palais, and then were put together in a book published in France by Belin in 1984, and in English: *The Beauty of Doing Mathematics*, Springer-Verlag, 1985.

PATRICIA CHWAT. I asked, very naively, if Mr. Lang could come to my school to talk to my students, and he agreed. He came last year, and this year he must have liked it, because he came back.

SERGE LANG. That's right, last year you had a really good class, an eighth grade class, and they all appeared very smart. It worked out so well that I thought I would come back this year, and I met an eighth and ninth grade class. It also went very well.

JEAN BRETTE. But the subjects you talked about were a little difficult. Usually, one does not give many proofs at the eighth or ninth grade level.

PATRICIA CHWAT. But one can, one can.

SERGE LANG. One can?

PATRICIA CHWAT. Not necessarily in the eighth grade, but in the ninth or tenth grade one can prove a lot in arithmetic. I believe this is the context which lends itself best to proofs. In geometry, they "see" things, and so they don't feel the need for proofs so much; also because previously they have made experiments, constructions, and it all seems clear to them. But when they see multiples, divisors, that's something new, and they feel that to be sure a result is true, for any number, because there are infinitely many of them, a proof is necessary and contributes to their understanding.

SERGE LANG. Do you do the proofs? Are they part of the program?

PATRICIA CHWAT. I don't know if they are officially on the program, but I certainly do them. The theorems, more or less, are on the program, and I do the proofs.

SERGE LANG. Yes, the theorems are on the program. I have seen books which follow the program, which have the theorems, but which don't have the proofs, whether in algebra or geometry. For instance, for the area of the disc or the volume of the ball, or the volume of a pyramid, they've got the formulas, but no proofs. That's what they call being "on the program".

PATRICIA CHWAT. But whatever the level, even at the university, you must accept some things without proof, things which you agree will be proved in later years. You need something to work with, right away.

SERGE LANG. Of course. I am not against granting some statements without proof, and some proofs can be given later, but the question is: "What is fundamental, and which proofs must be given, when?"

PATRICIA CHWAT. Yes, but if you manage to give the proof for the formula giving the volume of the ball in the eighth grade, I invite you to do so.

SERGE LANG. OK, I'll come.

PATRICIA CHWAT. In the eighth grade? Because that's when they see the formula for the first time.

SERGE LANG. Well, no. OK, I agree, you can give them the formula without proof in the eighth grade, and prove it in the ninth or tenth grade. But then it's not part of the program any more.

PATRICIA CHWAT. Rather cleverly, spatial geometry is part of the program in the eighth grade in physics, and in the ninth grade in math. Mathematicians think that children are not sufficiently mature to understand these notions in the eighth grade. The people responsible for making up the physics program decided otherwise. Then spatial geometry disappears, until the children are 16 years old. Hence they don't see proofs of these formulas at all.

SERGE LANG. But even in physics, they don't see the proofs.

PATRICIA CHWAT. That's true. I myself saw them only at the university.

SERGE LANG. There is the scandal! Those proofs are very beautiful, it's real mathematics. They allow you to appreciate mathematics, to see why something is true by using arguments which are quite understandable. I did it to students in the ninth, tenth and eleventh grade, and they understood, they got a lot of pleasure out of it. But it's not officially on the program, and it's not usually done at all.

PATRICIA CHWAT. I don't want to defend the programs systematically, but still you need to teach a large number of notions. What you did is not in the program, it's like folklore, it's very exceptional.

SERGE LANG. It's exceptional because it's not in the program, but it would be natural if it were on the program. All you have to do is make up programs where it would be natural to give such proofs.

JEAN BRETTE. Do you want to make one up?

SERGE LANG. No, I don't want to make up a program, but I wrote a geometry book.[2] I know what should go into a program. If it isn't there, that's not a law of nature, it's because the people who make up the programs have decided one way rather then another.

PATRICIA CHWAT. Of course, of course. One of the problems is that those who make up the programs are not actually teaching. As long as there will be "inspectors"[3] who don't teach, there won't be coherent programs.

---

[2] *Geometry, a High School Course*, in collaboration with Gene Murrow, Springer-Verlag, 1983.

[3] The French system of education is very centralized, and based in part on a system of "Inspectors" who make up programs and see to it that they are followed.

SERGE LANG. [*Laughing*] So you have to fight a bureaucracy, you have to fight a system, here as well as in the United States. But that's what one has to challenge, one has to try to do something else.

JEAN BRETTE. As far as you know, the situation is the same everywhere?

SERGE LANG. Yes, it's the same everywhere.

JEAN BRETTE. You mentioned one possible exception, I think? Where it was perhaps somewhat better?

SERGE LANG. I don't know exactly what happens in the Soviet Union. I have seen some books, elementary books, which seemed to me better than those which I saw here.

JEAN BRETTE. What do you mean by "elementary"? Eighth grade? Ninth grade?

SERGE LANG. Yes, and tenth grade. They seemed better, but I haven't seen enough of them to have a well documented judgment.

PATRICIA CHWAT. Anyway, books about mathematics right now are changing. Some of them are old-fashioned, and are similar to those which my teachers had when they were students, but others reflect a new spirit and force the student to think more, to raise questions. But when you can, there is nothing that prevents you from doing things which are not officially in the program. In my courses, I don't say that I spend most of my time doing things which are not in the program, but still, I do things at the edge, and try to do extra material, while trying to teach them whatever they have to know officially.

SERGE LANG. Not in the program . . . But you can do everything that's in the program in a coherent framework. That's what I object to, that's what drove me out of my mind last year, just before the conference at the Palais, to see a text for the 11th grade which was incoherent from beginning to end, with an accumulation of meaningless exercises. For instance, in geometry, you try to find the length of the circle, in the ninth grade, starting with the area of the disc. You need a formula which gives the expansion of $(r + h)^2$, what they call an identity. OK, you need it, and you have to do it anyhow, whether it's officially on the program or not. You need that $(r + h)^2$ is equal to $r^2 + 2rh + h^2$.

JEAN BRETTE. It's on the program for the 10th grade.

SERGE LANG. But you can do it before, what does it matter, do it whenever you need it. In the long run, you end up doing what is in the program, but you do it in a coherent framework, following a beautiful phrase, like a musical phrase, which hangs together. Instead of little things, one after the other, piled up without rhyme or reason. That's what I protest against, this accumulation of little things, which don't fit into a larger pattern which would help to remember them. [*Turning to Stephane, 14 years old.*] What do you say?

STEPHANE. I agree completely, because they give us exercises with numbers, but they don't mean anything. They give us numbers, and we have to add, subtract, that's all we have to do.

SERGE LANG. So you mean it's artificial?

STEPHANE. Yes. Besides, it's been two years that we have done almost nothing else, except a little geometry. Just checking that we know how to count.

PATRICIA CHWAT. But you must work out basic exercises to get acquainted with mathematical notions.

SERGE LANG. I never said you should eliminate basic exercises. The negation of one extreme is not the extreme of opposite type! Don't claim that I said what I did not. I don't want to eliminate all systematic exercises. You need those too, but the programs have eliminated the coherent stuff, the beautiful musical phrases in mathematics.

PATRICIA CHWAT. You can always put them back in.

SERGE LANG. Yes, you can put them in a book, but then some teachers will tell you that it isn't officially on the program, so they can't do them.

JEAN BRETTE. Or that they don't have the time to deal with them.

SERGE LANG. That's it, "I don't have the time to do it". But that's totally false, they could do it in addition to what's on the program.

JEAN BRETTE. [*To Patricia Chwat.*] And it isn't true, because you manage to do both.

PATRICIA CHWAT. Yes, because one needs to establish a certain continuity, and it bothers me that the students haven't seen certain things. But they need some continuity in what they learn, and to go from one grade to the next, they need to have acquired some basic knowledge which will be expected of them when they more from the eighth to the ninth grade.

SERGE LANG. But you can learn this basic knowledge in a coherent framework. I claim that to do basic material in a coherent framework is not incompatible. Whereas the official programs are designed as if there were some incompatibility. It's just not true. It reminds me of Stephane's notebook, which he showed me earlier. What did your teacher write?

STEPHANE. "Not in the program." I was not allowed to use a square root because it had not been officially discussed in class.

SERGE LANG. And you got a bad grade because of this?

STEPHANE. Yes.

SERGE LANG. Well, it's ridiculous, that's all.

PATRICIA CHWAT. He is not in my school. [*Laughter.*]

JEAN BRETTE. It's another school in the same district, but we won't say which one out of charity.

SERGE LANG. But it's scandalous! To give a bad grade to some kid because he did something very coherent, very intelligent, which showed he understood what was going on very well, but it just happened not to be on the program. It's insane to grade that way, but it happens. That's what disgusts kids from doing mathematics.

JEAN BRETTE. It depends more on the teacher than on the program, in this instance.

SERGE LANG. It depends on both. They are not mutually exclusive. It isn't because one is lousy that the other is good. The other could be good if they wanted, but they reinforce each other. Naturally, there are some teachers who don't know what they are doing and can't do better than what they do. There are many of those. There are also good ones, very good ones.

PATRICIA CHWAT. Now you raise the more general problem of training math teachers in France.

SERGE LANG. Not only in France. In the United States too, I know the situation better over there. In the teachers' colleges, for the most part in mathematics, it's a disaster.

PATRICIA CHWAT. In France, we have the IREM.

SERGE LANG. What's that?

JEAN BRETTE. Research Institutes for the Teaching of Mathematics, which were spread all over, and which are now dying because they are not properly funded any more. There aren't enough subsidies to allow teachers to attend the special courses, to allow them to teach less so they have time to learn new things.

PATRICIA CHWAT. They started when there was a big reform in high school teaching, and when the "new math" started. Now teachers are supposed to have been all recycled.

JEAN BRETTE. . . . and so the government thinks that the IREM don't serve any purpose any more.

SERGE LANG. That's another stupidity, "new math", so called "modern" math, or "old" math. Those categories don't exist. There is good math, and lousy math, which are not interesting. OK, in what they used to do before the "new math", there was also some incoherence, little things which did not mean anything. But they replaced old fashioned meaningless little things by new-fangled little things which still don't mean anything. So what's the improvement?

JEAN BRETTE. "Modern math" is not so new as you think. Here I have a book, an old book, dating back to 1746, where they refer to

Modern Mathematics, with a capital M for "Math" and "Modern". [4] Polemics on pedagogy don't seem to have changed much.

SERGE LANG. But there should not be polemics on words like that. It can't lead to anything constructive. You should not compare modern mathematics with ancient mathematics. That's not the proper approach. They are not opposed to each other. What are in opposition are coherent, beautiful mathematics, instructive and useful mathematics, with mathematics that aren't good for anything, or rather which are good only to turn off the kids. That's the opposition. Just using words like modern math or old math already prejudices the discussion along the wrong channels. I am not for Euclid or against Euclid. Take in Euclid whatever is still useful, and forget about a lot of it which isn't. Of course, Euclid piles up hundreds of pages, but today we can manage to cover what is important in thirty or forty pages, and that's it. You go on to something else which was discovered later in the history of mathematics. We have another perspective. What do you say, Stephane?

STEPHANE. I agree. Perhaps the new math is useful, but what's the point of doing it for a year or a year and a half?

SERGE LANG. What do you mean by the "new math"?

STEPHANE. Well, union, intersection, all of that. What you draw with circles.

SERGE LANG. There you are, some success! It shouldn't be done that way. It should just be an easy language, which is used and developed as you need it. There is no point making a big fuss about it.

PATRICIA CHWAT. There is some movement, today, to reduce the amount of vocabulary that students have to swallow.

SERGE LANG. I repeat: The vocabulary can be developed as you need it. If you spend weeks or months building up a vocabulary which is not used right away, but is an end in itself, then again it's a little thing. It's a "new" thing, but it's just as little, cut up in little slices, without nice "phrases", without beauty. There is nothing except an accumulation of little things. That is what should be eliminated. And it's not incompatible with learning basic notions and techniques. Your guy there, in the book from 1700, he saw it very well. What did he write?

JEAN BRETTE. Yes, for instance, he has a footnote here, which said just that:

> *Generally one should never give a definition to children without previously having shown them the thing that was being defined. The name should*

[4] *Institutions de Geometrie*, by M. De La Chapelle, Vol. I, 1746.

come only after the idea, because the name was used only to evoke the idea.

SERGE LANG. That's it.

JEAN BRETTE. Yes, and there are lots of other instructive comments in this book. The author wanted to persuade his contemporaries that one could teach mathematics to very young children, and he had given a great deal of thought on how to do this.

SERGE LANG. What's the guy's name?

JEAN BRETTE. M. de La Chapelle. I don't know much about him, but I think he was a friend of d'Alembert, and his book had the official approval of the Academy of Sciences. We don't have the time to read it now, but it might be worth while to publish some parts of it, in contemporary french.

SERGE LANG. Let me come back to a concrete case, because to be useful, a discussion must be based on concrete cases. I am not proposing a new ideology. If you put a label on something, you prejudice the discussion. Let's come back to the concrete case of $(r + h)^2$. Not all the kids know the formula for this, and it's like many things in mathematics. On one hand you need to understand theoretically what happens, and you must see the proof. On the other hand, you must also know how to use the formula automatically when you need it. There is no way to avoid this, and so you must ask the kids to repeat the formula ten times:

$$(r + h)^2 = r^2 + 2rh + h^2.$$

JEAN BRETTE. All together?

SERGE LANG. Definitely, all together, like a chorus. I make them do this. Do you?

PATRICIA CHWAT. Sometimes I make them recite something like that, but rarely. I don't make a habit of it.

SERGE LANG. When I was a kid, that's how I learned $(a + b)^2$, and I never forgot it. That's how my teacher taught it to me, and that's how it must be done. There is no other way. It must be driven into their ears like music. You shouldn't ask every time why the formula is true! You must be able to use it when you need it. And the only way–or rather: I never heard of another way–is to repeat the formula ten times, and again ten times before going to bed. I tell them: repeat the formula. At first they don't take it seriously, but they repeat it with me, like a chorus, and it amuses them. Stephane, that's the way I did it with you. What did I ask you to repeat?

STEPHANE. The volume of the ball.

SERGE LANG. Right, we had proved the formula, but you did not remember it at first. Then what did I make you do?

STEPHANE. Well, repeat 4/3 of πr^3.

SERGE LANG. Exactly. Two weeks ago. And now you will remember it till death do us part?

STEPHANE. Yes, I think so.

SERGE LANG. So, you must learn the formula by heart by repeating it ten times, but you must also see the proof. Because there are two different ways the brain works, but they are definitely not incompatible, they are complementary to each other. You must know both. And it's the same thing at Yale, with university students who are 17 and 18 years old. Some of them, too many, are not able to repeat this formula automatically when they come from high school, and I ask them. Maybe a third, or half the class. It's a scandal!

JEAN BRETTE. It isn't worse than here.

SERGE LANG. It's the same, no doubt about it. And even though they are older, like 17 or 18, I still make them repeat the formula ten times. [*Laughter.*] First they think I have it in for them, but after a few days, they end up by understanding that I am right; that it's an effective way to learn, and that they have to use it on certain occasions. But they must also see the proof, and understand how to formulate it. They need both. Afterward, they make jokes about the fact that I make them repeat ten times the same thing.

JEAN BRETTE. You have gone through several experiences of high school teaching, around Paris as well as in Canada, at Toronto. How do you appreciate the students' reaction to what you do, which is usually not in their standard program, or what they usually do?

SERGE LANG. You have seen how the students react.

JEAN BRETTE. Yes, we have seen some examples, but you, how do you appreciate it?

SERGE LANG. It's a pleasure for me! They open up, they react positively, they are immediately interested, and they start taking notes. After that, I think you asked them . . .

PATRICIA CHWAT. Yes. They have a special homework every two weeks, and the other day, I asked them to write up what you did on the length of the circumference of the circle.

SERGE LANG. I saw their homework, which you brought today, and many of them were able to reproduce everything. You did not help them?

PATRICIA CHWAT. No. Of course, not all of them remembered everything, not all the details, but almost all of them remembered the general idea for the proof.

SERGE LANG. I saw how they wrote it up. It was wonderful. It was neat, coherent, just great.

PATRICIA CHWAT. They certainly were neat.

SERGE LANG. It was not only neat, it was also coherent. There were sentences, explanations, well thought out expositions. They learned certain formulas, so they learned some algebra, they learned a proof, and they learned how to use language to express their mathematical thoughts. That must come at the end. When we are doing math together, at first I start a sentence, and I ask them to complete it with one or two words, or to finish a sentence which I myself began. But as we go further along, I ask them to create their own sentences, to explain their own mathematics. Then at some point, I reverse the process, and I ask the students to express themselves what's in their heads, and to extract by themselves their own mathematical thoughts from their own head. It's for the student to start his own sentence. You saw it with Yaelle, at the end, when I asked her to repeat the proof which gives the length of the circle, $c = 2\pi r$. That's exactly what she did. With her hands, with her eyes, with her brain, she did it.

JEAN BRETTE. We almost heard her think.

SERGE LANG. We *saw* her think. The expression on her face, her concentration . . .

PATRICIA CHWAT. It's true, I have gotten them to talk. We haven't mentioned this yet, but in my classes, they are used to speaking up, finishing sentences, and then making up their own sentences. In your case, it wasn't the same person in front of them, but it was not a technique fundamentally different from what they often see in class.

SERGE LANG. I know university students who are not able to do that in mathematics. Maybe half, in an average class, can't do it. Of course, there are some honour classes, brilliant classes, but in an average class in the United States . . . all they are required to do is to fill out a little square. They are given a problem, with a numerical answer, and all they are asked to do is to put a number in a little box. That they usually can do well. But to create their own sentences, to explain by themselves what's going on, to write up a proof, they can't do it. It's a scandal! And there are many classes, here in France–I don't mean yours, because you do better than that [*Laughter*]–what about in your own class? Stephane?

STEPHANE. It's the same thing. Usually they just ask for a result. Well, sometimes they ask us to write something up, but not often.

JEAN BRETTE. One shouldn't wonder that students can't write things up if their teachers never ask them to learn proofs.

SERGE LANG. Of course, but it helps to clarify ideas, to write them up, and the subject becomes much more interesting for students if one asks them to furnish this effort, which anyway is not so great. It gives them pleasure. Listen, when Yaelle repeated the whole proof, all at once, you saw what happened, the whole class applauded spontaneously. [p. 89, *The length of the circle.*]

PATRICIA CHWAT. But you happened to pick on Yaelle, it would not have gone exactly the same way for every student in the class. It was clear it would give her pleasure. Others would not have taken it so well.

JEAN BRETTE. What do you mean, being picked on?

PATRICIA CHWAT. Yes, to redo the whole proof.

JEAN BRETTE. But there is also the example of Rachel, in Toronto.

SERGE LANG. That was another case. In Toronto, there was one student who would not have been able to do what Yaelle did, reproduce the whole proof, and I had to lead her through it step by step. That's true, mathematical talents differ, and all students are not able to do it, the first time, but if I met them again for another hour the following week, many would become able to do it.

PATRICIA CHWAT. But it's a question of time, and if we had enough time, we could retrieve practically everybody, except in pathological cases. But we have 25 students, sometimes more; some students are already very much behind, and unfortunately, we have to leave them like that. It's a dramatic situation, but . . .

JEAN BRETTE. I agree, but on the other hand, an exchange like the one with Yaelle was valuable for the whole class, because there was a certain psychological tension, there was suspense, the others were waiting to see if she would succeed in reconstructing everything, and they were following along at about the same speed. It benefited everybody.

PATRICIA CHWAT. But some of them were quite glad that they were not picked on, and probably were thinking about something else, you have to be honest.

JEAN BRETTE. Possibly, I am not saying that everybody was following, but an exchange with so much tension, so much concentration . . .

PATRICIA CHWAT. For sure, Yaelle was carried by the class, she was encouraged, but some of the others were very glad to be doing only that.

SERGE LANG. I would have handled another student differently, I would adjust to the ability of the student, if I picked someone else. I don't do exactly the same thing with different students. If I realized that a student was less able then Yaelle to repeat the proof, I would have handled the discussion differently, just as I did it with Rachel in Toronto: first I had to lead her step by step to repeat the proof, but she ended up by saying more than she knew in the beginning. I pick on them at random. I don't take those who raise their hand, I pick on those who don't.

PATRICIA CHWAT. There is a problem with that. One of my students was very disappointed because she had made an effort to raise her hand, she is rather shy, and when she raised her hand, you did not ask her.

JEAN BRETTE. She was in the first row?

PATRICIA CHWAT. She sat next to Yaelle. She is not very gifted in mathematics, but in this case she wanted to show you that she could do something, and you did not ask her just because of that. [*Laughter.*]

SERGE LANG. Yes, OK, there are risks, I don't know the students, and I see them only once . . . What I wanted to avoid was that only those who usually know what's going on should answer. It's very easy to fall into this trap. All situations can arise. There are those who are not able to repeat the proof by themselves, and there are those who are able, even if they don't think so. All cases are possible. What is needed is to bring them all to the maximal level of understanding.

PATRICIA CHWAT. And that requires time.

SERGE LANG. Yes, it requires time, but it's a long term investment. Once they have learned how to do it in one case, they will do it much more easily in the next case. The time is not lost.

JEAN BRETTE. This goes back to what you were saying a while ago. If you eliminate from the programs what you think is unnecessary, then it gives time to do the rest.

SERGE LANG. Exactly, that's right. If you clear the deadwood from the programs, and if you construct them coherently, then it leaves plenty of time to do beautiful mathematics, which will be useful, with all the necessary proofs, and whatever formulas you have to know automatically, just like that.

JEAN BRETTE. Stephane, you were sitting in the back of the class, as a guest. What do you think?

STEPHANE. It was OK, but perhaps if it happened all the time, one might get tired of it, the students might be less awake, in the long run.

SERGE LANG. Oh you know, at the university, even with 17 year olds, I know how to wake them up. It suffices to do this [*pointing the finger*] and I guarantee you that it wakes them up. Besides, they never know when it's going to hit them. [*Laughter.*] It's not a big problem to keep them awake.

Besides, I can keep them awake by showing them real mathematical problems.

PATRICIA CHWAT. But in the 8th or 9th grade, they're too young. Students in the 8th or 9th grade know almost no mathematics.

SERGE LANG. Yes, but still there are certain topics . . . when do you do prime numbers, in what class?

PATRICIA CHWAT. In the 8th grade.

SERGE LANG. Then in the 9th grade, they are a little more mature.

PATRICIA CHWAT. Yes, but then we don't talk about this any more.

SERGE LANG. You don't talk about it any more? Here we come back to the same problem we met before. You state a result in the 8th grade, and after that, it's over, you don't follow it up.

PATRICIA CHWAT. You use prime numbers just to write rational numbers in the 9th grade. You have to know simple fractions, but conceptually, there is no progress. You use notions which were introduced in the 8th grade.

SERGE LANG. All right, but I can tell you a problem about prime numbers which could be discussed in the 8th grade, just when you define prime numbers. First you can ask them if there is a infinity of them, and you can prove it. At this point, you tell them: look at the twin primes, 3, 5; 5, 7; 11, 13; and the next one, Stephane?

STEPHANE. 17, 19.

SERGE LANG. Right. And you can raise the problem of the twin primes, are there infinitely many? And you see how they react. They can have fun with this, and find a number of them, maybe up to 100. It will make them work with integers, addition, subtractions, divisions. But they will do so in a meaningful context, to answer a question. After they have computed like that, up to 100, they will observe that the list of twin primes seems to go on, and you ask them: "Is there an infinite number of them?" Then you'll see what they answer. Some will say yes, some will say no, then you ask them why, for what reasons, what is their intuition, and after a while, the same day maybe, you let them go home and compute more of them, so that they arrive at a more sophisticated conclusion. And a week later, after some suspense–and there is suspense: Is there or is there not an infinity of twin primes–you tell them: "Nobody knows the answer! And if you succeed to prove it, you make it in the history of mathematics."

PATRICIA CHWAT. Yes, but to raise this kind of problem, there is one necessary condition, you have to know them. You need a certain mathematical culture.

SERGE LANG. But for god's sake, what's the use of teacher's colleges, which are supposed to train the teachers?

PATRICIA CHWAT. But there aren't any!

SERGE LANG. So it's for society to figure out how to give this training. Instead of putting all sorts of stupid things in their books, they should put this in. It will make the students work and think. They will learn to divide just as well, they will get trained in the basics just as well, but starting from problems which have some meaning, which are coherent, and quite beautiful! People react very positively to such a problem. And it's the same thing for some equations. When you do x and y, you can do what I did in my lecture at the Palais de la Découverte[5] or in Toronto.

PATRICIA CHWAT. By the way, some of my students got a reprint of the talk which you gave last year, and on the front page, in big type you have the equation $y^2 = x^3 + 1$. During my next math class, Yaelle asked me what is an equation. I had not yet done this (even now, I have not done it in class), but I explained this mathematical term to her, that is the notion of an equation; and Yaelle told me: "Ah, that's an equation, so it's very easy, and I thought it was something very complicated." And Nathalie, the one you did not want to question, tried to find solutions, and she said: "Well, let $y = 3$ and $x = 2$."

SERGE LANG. There you are, I have nothing to add. [*Laughter.*]

JEAN BRETTE. And then, did she read the whole conference?

PATRICIA CHWAT. I don't know, I did not ask her, but both were very disappointed to see that in the next homework, there was no equation to solve. I told them that the problem did not consist only in finding solutions, but also in showing that there were no others.

JEAN BRETTE. But you have to start by finding some, there are two problems.

PATRICIA CHWAT. What I mean is that for them, it was a game.

SERGE LANG. Absolutely. That's a marvelous reaction, because if a child reacts that way, then she is caught by the problem, and then you can develop it and learn even what's officially on the program, but in the context of this problem.

PATRICIA CHWAT. Yes, but one has to find contexts which are sufficiently rich to allow you to encompass . . .

SERGE LANG. There will be algebra, there will be some geometry, there will be coordinates, the whole thing will come together. But you are then motivated by a larger structure, a musical phrase, a mathematical phrase, which is a beautiful phrase . . .

JEAN BRETTE. You really like music!

SERGE LANG. Definitely, I like music. Yes, because I have to communicate, and if I say "music", it stimulates people. They are not used to think in similar terms for music and mathematics. But not in a trivial way, not by counting seconds, or the number of vibrations. That's not what I mean by a musical phrase and its mathematical analogue, the mathematical theme which can be developed.

Mathematics is not just "numbers", any more than music is notes. Music, that's what happens when you enjoy it, when you get a high out of

[5] *The Beauty of Doing Mathematics*, Springer-Verlag, 1985.

it. You can get a high with mathematics. You just said so, when Nathalie found a solution for her equation, she was very happy.

PATRICIA CHWAT. Me too, naturally.

JEAN BRETTE. Yes, and what's nice is that this conference, you can read a good half of it before you reach a point which you don't understand, which will give you trouble.

PATRICIA CHWAT. Last year, I had used your reprint on prime numbers[5] and we had looked at it during the last class, as a reward, and we had gone quite deeply into it. Of course, we had some trouble with logarithms, because in the 8th grade, you know, it's tough.

SERGE LANG. Well, naturally.

PATRICIA CHWAT. In any case, this talk was not addressed to 8th grade students.

JEAN BRETTE. It wasn't addressed to anyone in particular. It was addressed to those who were there, and if at some point they stop understanding, they have already gone that far, and what's more, they know that it's not the end of it. It's true that rarely if ever in elementary teaching does one tell the students that mathematics is not something which is finite, which was invented two thousand years ago, or two centuries ago, or fifty years ago. It makes people feel good to know that great mathematicians also don't know certain things and that there are problems which nobody knows how to solve.

PATRICIA CHWAT. It's good for other people's morale!

SERGE LANG. For one thing, but for another, I said it explicitly in one of the classes, when I told them that all those theorems, someone had to discover them, maybe a thousand years ago, or two hundred years ago, and then one of the kids said: "Well, then, what you discover today, it will be taught to us in a class in two hundred years." Do you remember? So they understood the human side of this thing, because one of them asked what I did in life, and I said that I do research, I discover theorems. And she said: "And what happens if you can't discover theorems any more?" And one of the kids behind her shouted: "Resign, Resign!" [*Laughter.*] They sure got the idea, they understood, and mathematics became alive for them.

PATRICIA CHWAT. I tell them that mathematics exists, that mathematics is alive, and not stereotyped, old and musty, but it's not always easy with very young children.

SERGE LANG. You are right. Each teacher must do according to his own way, his own taste. Each one must use their own means to excite the students. One needs everything, without exclusivity.

JEAN BRETTE. Here is a perfect conclusion: One needs everything, without exclusivity.

9 780387 961293